CHECKPOINTS 2002

VCE PHYSICS 1

Past exam questions from 1995–2001

Over 300 questions with answers

Summary notes for Unit 3 theory

Syd Boydell & Russell Tytler

CAMBRIDGE UNIVERSITY PRESS

PUBLISHED BY THE PRESS SYNDICATE OF THE UNIVERSITY OF CAMBRIDGE
The Pitt Building, Trumpington Street, Cambridge, United Kingdom

CAMBRIDGE UNIVERSITY PRESS
The Edinburgh Building, Cambridge CB2 2RU, UK
40 West 20th Street, New York, NY 10011—4211, USA
477 Williamstown Road, Port Melbourne, 3207, Australia
Ruiz de Alarcó n 13, 28014 Madrid, Spain
Dock House, The Waterfront, Cape Town 8001, South Africa

http://www.cambridge.edu.au

© Cambridge University Press 2001

Notice to teachers
It is illegal to reproduce any part of this work in material form
(including photocopying and electronic storage)
except under the following circumstances:
(i) where you are abiding by a licence granted to your school or
institution by the Copyright Agency Limited that permits the
copying of small parts of text, in limited quantities, within the
conditions set out in the licence;
(ii) where no such licence exists, or where you wish to exceed the
terms of the licence, and you have gained the written permission
of Cambridge University Press.

First published in 1995 by Coghill Publishing Pty Ltd
Reprinted with changes 1998
Second edition 1999
Third edition 2000
Fourth edition 2001
Fifth edition 2001

Printed in Australia by Brown Prior Anderson
Typeface Times 12pt

National Library of Australia Cataloguing in Publication data
    Boydell, Sydney
    VCE physics 1, 2002
    New ed.
    ISBN 0 521 61089 3 (Series)
    ISBN 0 521 01289 9
    1. Physics - Problems, exercises, etc. 2. Physics - Examinations, questions, etc. 3.
    Victorian Certificate of Education examination - Study guides. I.Tytler, Russell.
    II.Title. III. Title: Victorian Certificate of Education physics (Series: Cambridge
    checkpoints).
530.076

ISBN 0 521 01289 9 paperback

Disclaimer:
The text is independently published for use by VCE students by Cambridge University Press. The text is in no way connected with, or endorsed by, the Victorian Curriculum and Assessment Authority or the Ministry of Education in Victoria.
Teachers and students are advised that the solutions are suggestions only, and are reminded that in order to satisfy the requirements of school and external assessments, candidates must submit work that is clearly their own.
The contents of this book follow the assessment criteria for the 2002 examination tasks as published by the Victorian Curriculum and Assessment Authority. Students and teachers should be aware that the criteria may change from year to year at the discretion of the Victorian Curriculum and Assessment Authority.

ACKNOWLEDGEMENT: Past exam material reproduced with permission of the
Victorian Curriculum and Assessment Authority

PREFACE

This book is a collection of over 570 practice/review questions for the VCE Physics written examination held in June. This examination covers content from the fields of Sound, Electric Power and Electronics.

The book includes all the relevant 'official' questions set for CAT 1 from 1996 through to the written examination in June 2001. They are clearly labelled as *(1996 CAT)* or *(2000 written examination)* and so on. Questions on material now outside the study design have also been omitted or modified.

In addition to the 'official' questions there are around 300 original questions in the same range of formats and styles, covering all the concepts in the current study design.

The questions have been grouped in thirteen chapters. The chapter titles do not correspond exactly to study design labels, but have been grouped for convenient revision.

This book should serve primarily as a thorough source of practice questions to be tackled after the relevant theory has been completed.

Answers and outline solutions have been provided to all questions.

Marks have been attached to some of the questions. This is to indicate the weighting such a question would have in the examination. Questions *without* marks attached are worth 2 marks.

The book also contains a point form summary of the complete course, grouped under headings which correspond exactly to the grouping of the review questions.

CONTENTS

Introduction	Tackling the examination	1
Chapter 1	Sound basics	3
Chapter 2	Sound intensity and level	16
Chapter 3	Standing waves and resonance	27
Chapter 4	Diffraction and interference	38
Chapter 5	Magnetic and electric basics	50
Chapter 6	Motors	59
Chapter 7	Generation principles	71
Chapter 8	Transformers, transmission and consumption	80
Chapter 9	Electronics basics	94
Chapter 10	Capacitors and diodes	106
Chapter 11	Amplification	117
Chapter 12	Logic	127
Chapter 13	Flip-flops	137
Appendix A	Sound summary	141
Appendix B	Electric power summary	150
Appendix C	Electronics summary	160
	Answers	170

INTRODUCTION   **Tackling the examination**

What to expect

The written examination is a test on three areas of study. Details of the current relative weightings given by VBOS are shown in the table below. Note that they are equally weighted. The suggested time allocation is one minute for each mark.

Area of Study	Marks	Suggested time allocation
Sound	30	30 minutes
Electric Power	30	30 minutes
Electronics	30	30 minutes
TOTAL	90	90 minutes
(Reading time)	0	15 minutes

You can expect a mix of questions. These will include multiple choice, numerical answer (just write a number in a box), 'explain' questions, 'sketch a graph' questions, 'label a diagram' questions, 'give an example' questions and others of this general kind.

Most of the questions will be worth 1 or 2 marks. Questions worth more than 4 marks are rather rare.

Preparation for the examination

The best preparation is always lots of practice at the types of questions you will meet in the CAT. That is the major aim of this book.

Many students, however, are uncomfortable doing practice problems before they feel they have a thorough grasp of the underlying theory. They spend hours and hours on notes before attempting the problems. Some even leave the problems until the very last. *This is a real trap!*

Doing problems (and sometimes getting them wrong) is generally the *best* pathway to a secure grasp of the theory. Notes (or summaries) and problems must go hand in hand. In the end, ability to do problems is much more important than a beautiful set of notes.

A good approach is to *modify the notes* as you discover things about your understanding from the problems.

If you get a wrong answer, it's often a good idea to try to rework the problem *before* looking at the outline solution.

1

The formula sheet and your own notes (your 'cheat' sheet)

You will find a list of formulas included with the examination paper. It contains the central formulas – but it is *not* a complete list of all the ideas or quantitative relationships.

The best approach is to ignore it and concentrate your efforts on building up a really good set of your own pre-written notes that you are allowed to bring into the examination. You are allowed up to two sides of A4 paper.

Once again, the best way to develop these notes is *at the same time* that you are doing practice problems. This will ensure that the material you write down is relevant. Hand-written notes (perhaps with colour) seem to work better for most students than computer-produced masterpieces. They also allow a freer use of diagrams – an essential element.

Tackling 'explain' questions

One way of approaching these can be summarised as follows:
- start with a diagram/graph/circuit/sketch
- make your explanation as a series of dot points
- quote key formulas whenever possible
- give 'typical' values of quantities whenever possible

The advantage of the introductory diagram is that it can help you organise your thoughts in the notes that follow it. Most markers love diagrams! Sometimes you will gain credit for a diagram even without any accompanying text.

It is not necessary to write in complete sentences, and some students find that the discipline imposed by dot points helps them organise their thoughts. However, if you do use complete sentences, keep the sentences as short as possible.

Using reading time effectively

The best use of this is to start thinking hard about some of the questions. The best ones to look at are probably the ones with long stems or lots of information to digest. A '4 mark' question can often take longer than its time equivalent of 4 minutes, so using your reading time to unravel the stem makes good sense.

Using working time effectively

- Don't use 'white-out' – cross out.
- Watch the time – force yourself to go on and come back later – don't get trapped by questions.
- Make a list of things to check at the end – analyse your weaknesses *before* you go in to the examination.
- Never leave a multiple choice unanswered – if you have to, guess between the most likely answers.

CHAPTER 1 — Sound basics

Question 1
The wavelength of a 340 Hz note is:

A	The distance between a compression and a rarefaction.
B	The distance between two adjacent rarefactions.
C	The time taken for a compression to travel a complete cycle.
D	The position of the main compressions.
E	None of the above.

Give a reason for your choice of answer.

[3 marks]

The diagram below models air molecules at an instant when a sound wave is passing through it. It is travelling with a speed of 340 m s^{-1}.

direction of travel of the wave

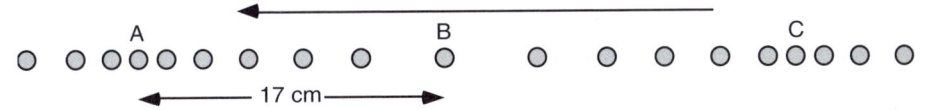

A   B   C
o o oooo o o o   o   o   o   o   o ooo o o

←——— 17 cm ———→

Question 2
Calculate the wavelength of the sound wave.

Question 3
Calculate the frequency of the sound wave.

Question 4
Which *one or more* best describes the pressure at points A, B and C?

A	It is higher at A and C than local atmospheric pressure.
B	It is higher at B than local atmospheric pressure.
C	It is lower at A and C than local atmospheric pressure.
D	It is lower at B than local atmospheric pressure.

[1 mark]

Question 5
Sound is often described as a *longitudinal* wave. Other waves are described as *transverse* waves. Outline the essential difference between these two waves. Use a diagram if you wish.

[4 marks]

The pressure variation caused by the sound of a single frequency from a loudspeaker can be described by the graph shown following. This graph shows the variation from normal atmospheric pressure at a specific instant of time.

1 Sound basics

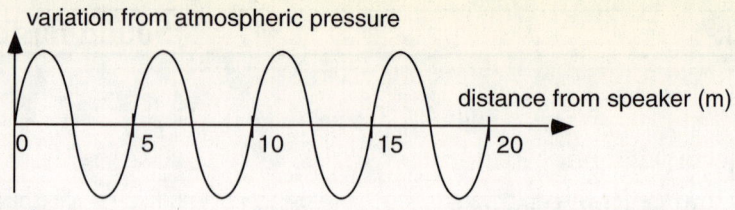

[Graph: variation from atmospheric pressure vs distance from speaker (m), sinusoidal wave with wavelength 5 m from 0 to 20 m]

Question 6
Graph this pressure variation three-quarters of a period later.

Question 7
What is the wavelength and frequency of this sound? (Take the speed of sound as 330 m s^{-1}.)

[3 marks]

Sound is emitted from a loudspeaker. About 1 metre from the speaker, a very light spider web is floating freely in the still air as shown in the diagram.

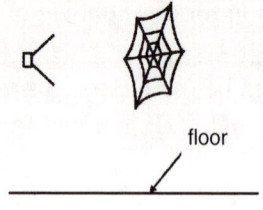

Question 8 *(1994 CAT)*
Which of the graphs below best represents the motion of any point on the spider web as a function of time? Note that the labels on the vertical axes of the graphs are different.

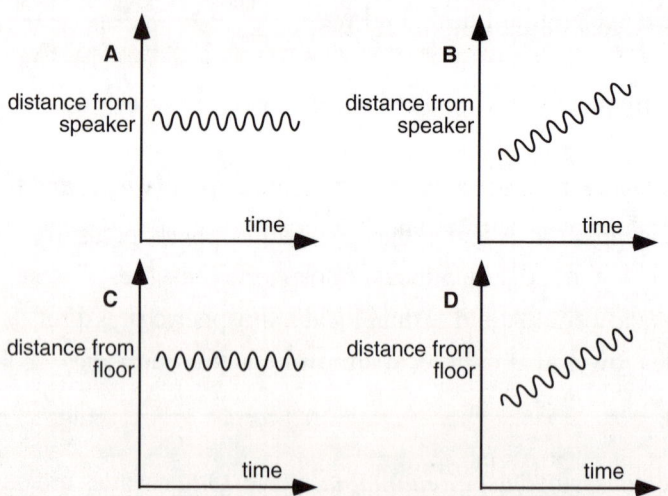

[Four graphs A, B, C, D showing different wave patterns vs time, with vertical axes labelled "distance from speaker" (A, B) and "distance from floor" (C, D)]

Question 9
Human ears in their prime can hear from about 20 Hz through to about 20 000 Hz. Give an estimate of the ratio:

$$\frac{\text{longest wavelength sound a human ear can hear}}{\text{shortest wavelength sound a human ear can hear}} ?$$

1 Sound basics

Question 10
When two sound waves meet, they:

A   pass through each other without being changed.
B   reflect off each other.
C   pass through each other, combining their effects when they overlap.
D   diffract off each other.

Give an example of a situation that illustrates your choice of answer.

[3 marks]

A single note from a large concert organ is being tested in a hall. A microphone placed near it produces the graph shown below on a CRO.

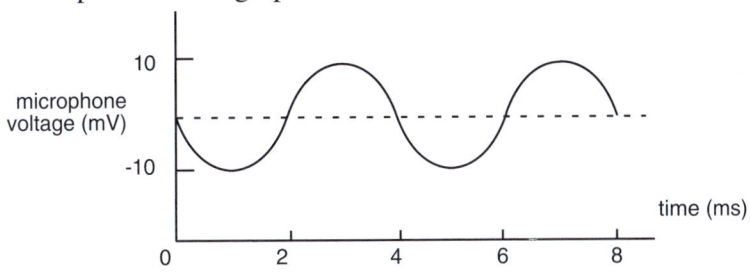

The microphone voltage is directly proportional to the variation in air pressure produced by the sound wave from the organ pipe.

Question 11
What is the frequency of the sound produced by the organ pipe?

Question 12
If the speed of sound in the hall is 345 m s^{-1}, calculate the wavelength of the sound wave produced by the organ pipe.

Sound from an organ pipe travels out as a *travelling* wave. The sketch following shows the approximate distribution of molecules between two points X and Y in the hall at a particular time.

Question 13
Sketch what the molecule distribution would look like *half a period* later.

A loudspeaker is being driven by an electronic instrument, producing a single frequency sound of 512 Hz. Freda is listening to the sound on the other side of the room. The speed of sound in the room is 336 m s^{-1}.

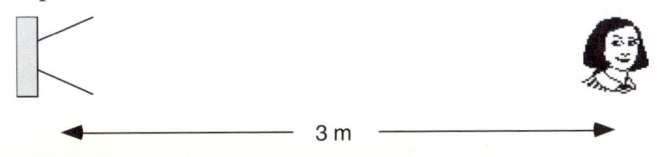

Question 14
What is the wavelength of the sound that Freda hears?

Question 15
How long does it take the sound wave to travel 3 m from the loudspeaker to Freda?

Question 16
Freda is some distance from the loudspeaker. Outline briefly what happens in the air to enable sound to travel from the loudspeaker to Freda.

[4 marks]

Question 17
You are standing beside a high brick wall at the point X in the diagram, listening to the band. To your surprise, there seem to be two bands, one where you can see it, and another one at Y. Explain. [4 marks]

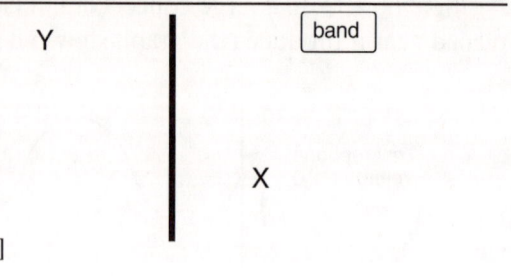

Layers of bedrock are sometimes discovered by setting off explosions at ground level, and measuring the time until a reflected wave appears at the surface, as shown in the diagram to the right.

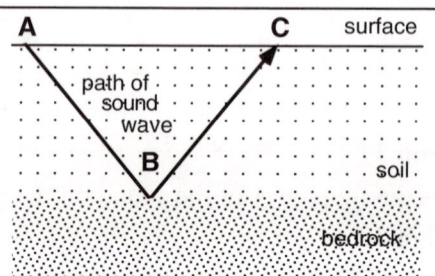

The explosion is set off at A, and the time for it to travel to C is measured. The *horizontal* distance between A and B is 340 m. The time taken for the travel of the wave from A to C is 1.6 sec.

Question 18
What is the horizontal distance between A and C?

Question 19
The distance AB is 480 m. Calculate the vertical distance between the surface and the top of the bedrock (the depth of the soil).

Question 20
If the temperature of the soil were to rise significantly, which *one or more* of the following best describes the effect on the above situation?

A   The angle ABC would increase.
B   The angle ABC would decrease.
C   The measured time interval would decrease.
D   The measured time interval would increase.

[1 mark]

1 Sound basics

Jennie measures the speed of sound using echoes. She stands 120 m from a large wall which reflects sound effectively. When she claps at a frequency of 1.60 Hz, the echo from each clap coincides with the next clap.

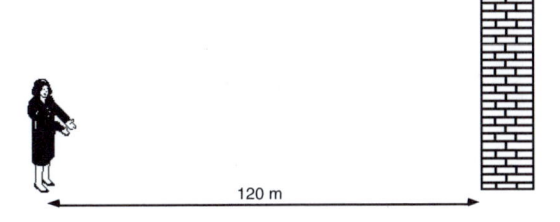

Question 21
What is the time interval between each clap?

Question 22
Calculate the speed of sound from this data.

Question 23
Jennie now claps at exactly *half* the frequency (i.e. at 0.8 Hz). Describe what echoes she hears and their timing. [3 marks]

Question 24
Jennie now exactly doubles her clapping speed, to 3.2 Hz. Describe what she now hears. [3 marks]

Question 25
If the speed of sound were to *rise*, which *one or more* of the following would keep the echoes synchronised with her original clapping?

A   She would have to clap faster.
B   She would have to clap slower.
C   She should clap at the same rate.
D   She should move towards the wall.
E   She should move away from the wall. [1 mark]

Rob is making measurements of the waveforms of single drumbeats. The drummer is standing as shown in the left-hand diagram below. The waveform for a single drum beat is shown in the right-hand diagram. To Rob's surprise the CRO shows *two* pulses on the screen.

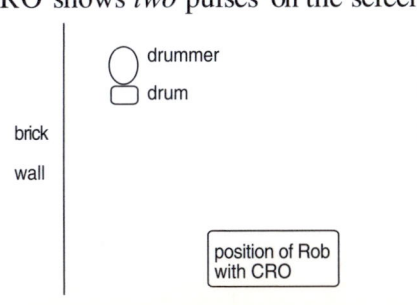

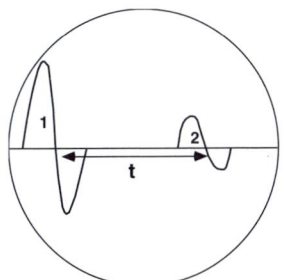

1 Sound basics

Question 26
Explain why there are two pulses on the screen, and why pulse **2** is smaller than pulse **1**, and what the time *t* represents.

[3 marks]

A flute produces a single frequency sound. A microphone near the flute, connected to a CRO, shows the air pressure varying with time as shown.

Question 27   *(1996 CAT)*
What is the frequency of the sound produced by the flute?

Question 28   *(1996 CAT)*
What is the speed of sound in the hall? (The wavelength is 0.70 m.)

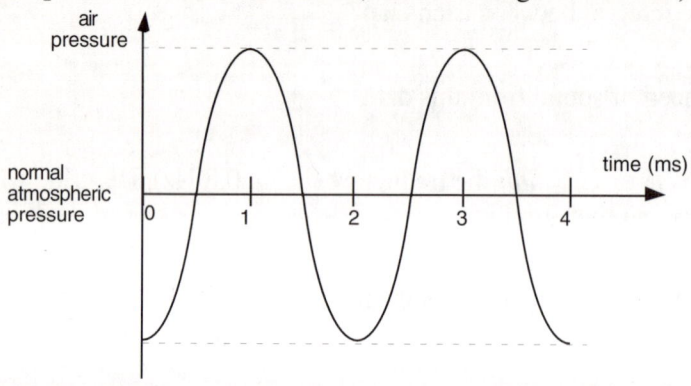

Flutes can be modelled as tubes open at both ends. The standing wave in the tube, shown in the diagram following, is at an instant when the flute is playing the single-frequency sound described above.

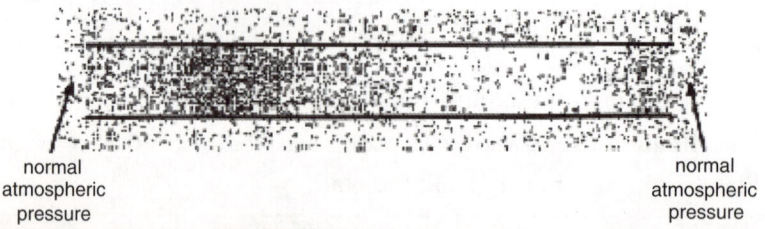

Question 29   *(1996 CAT)*
Draw a graph showing how the pressure depends on position along the tube at the instant shown in the diagram above.

Question 30   *(1996 CAT)*
What is the length of the tube shown in the diagram above?

Sound from the flute leaves it as a *travelling wave*. The diagram shows the distribution of molecules between **P** & **Q**. The wave is travelling to the *right*. The rectangle *X* contains a specific group of air molecules.

Question 31   *(1996 CAT)*

Which diagram (A–E) below best shows the distribution of air molecules between P and Q at a time *one period later* in the original rectangle X?

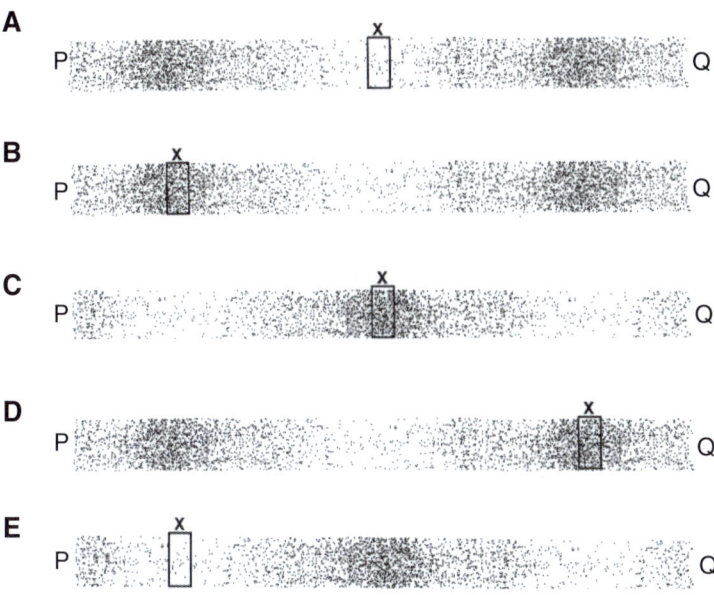

A

B

C

D

E

Sound barriers for a freeway aim to reduce the sound level nearby. Two designs do this by *reflecting* sound. The barriers don't *absorb* much sound. Barriers with large inclined front surfaces, shown following, reduce the sound level more effectively than those with vertical surfaces, also shown.

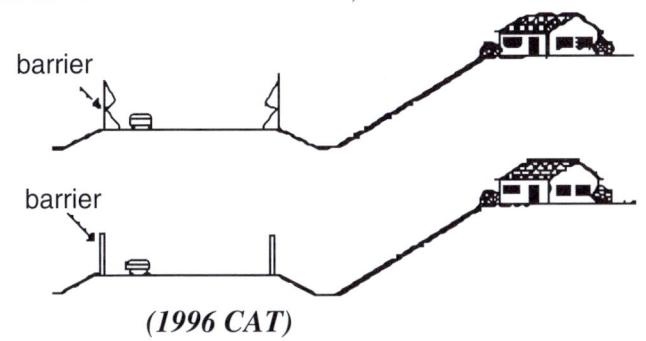

Question 32   *(1996 CAT)*

Sketch on the diagrams to show why this is so, and explain the physics.

[4 marks]

The sound barriers are found to be much more effective in reducing the sound level of a 5000-Hz siren than a 15 Hz sound from a truck.

Question 33   *(1996 CAT)*

Explain why the barriers are more effective in reducing the sound level of the siren.

[4 marks]

1 Sound basics

A cardboard cone loudspeaker acts as a
diaphragm moving backwards and
forwards, producing a travelling sound
wave in air. Such a speaker is mounted on a
wall as shown to the right. The dots
represent fine dust particles, floating at rest.
The graph shows horizontal displacement
of the speaker against time.

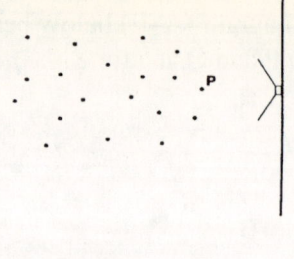

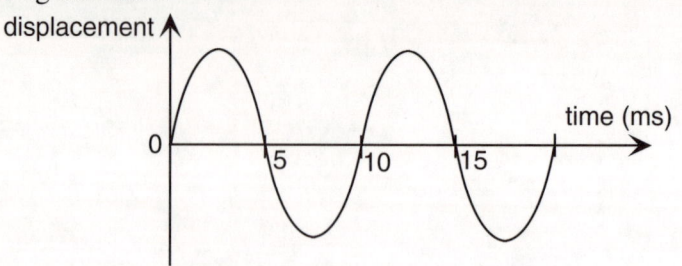

Question 34   *(1997 CAT)*

What is the frequency of the sound?

In the air are fine dust particles, floating at rest. They are shown as dots in the diagram earlier. When the loudspeaker is turned on, they are forced to move by the pressure variations associated with the sound wave.

Question 35   *(1997 CAT)*

For the dust particle at P, directly in front of the loudspeaker, which of the statements below best describes its motion? Justify your choice.

A   It vibrates vertically up and down at the frequency of the sound wave.
B   It vibrates horizontally backwards and forwards at the frequency of the sound wave.
C   It moves horizontally forwards, travelling with the sound wave.
D   It remains at rest.

[3 marks]

Students are asked to demonstrate how sound travels. Tina's group shows the class a long piece of flexible rubber hose and creates a wave by shaking it on the floor. Georgio's group finds a slinky and shakes it back and forward along the direction of its length. The two group's waves are shown below, against the floor laid out in square tiles of side 30 cm.

1 Sound basics

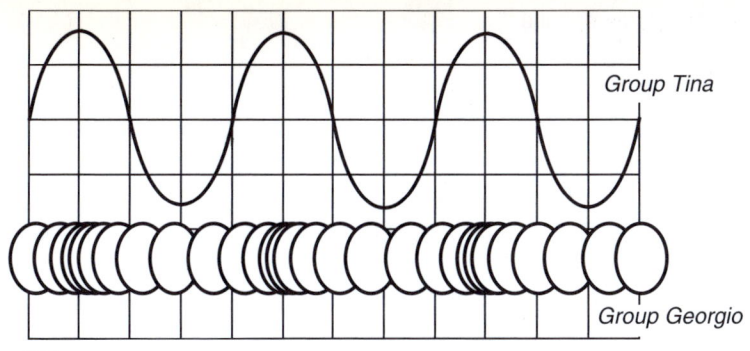

Group Tina

Group Georgio

Question 36
Estimate the wavelength of both waves.

Question 37
Write a brief explanation as to which model provides the better representation of the properties of sound waves.

[3 marks]

Question 38
Which of the following could represent the wave *one quarter* of a cycle later? The wave is travelling to the right. (Tina's groups' wave is shown in the graph above at t = 0.)

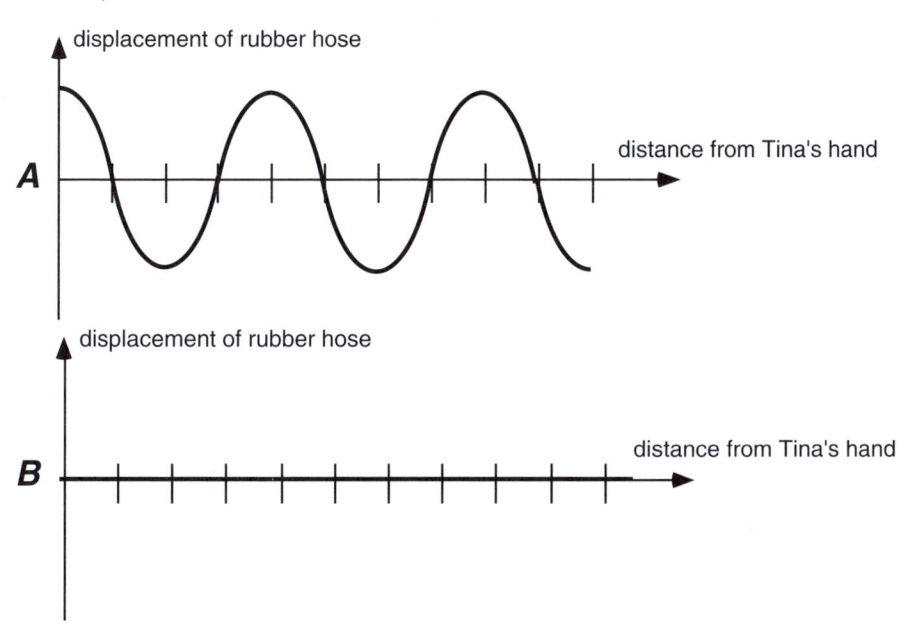

A
displacement of rubber hose
distance from Tina's hand

B
displacement of rubber hose
distance from Tina's hand

1 Sound basics

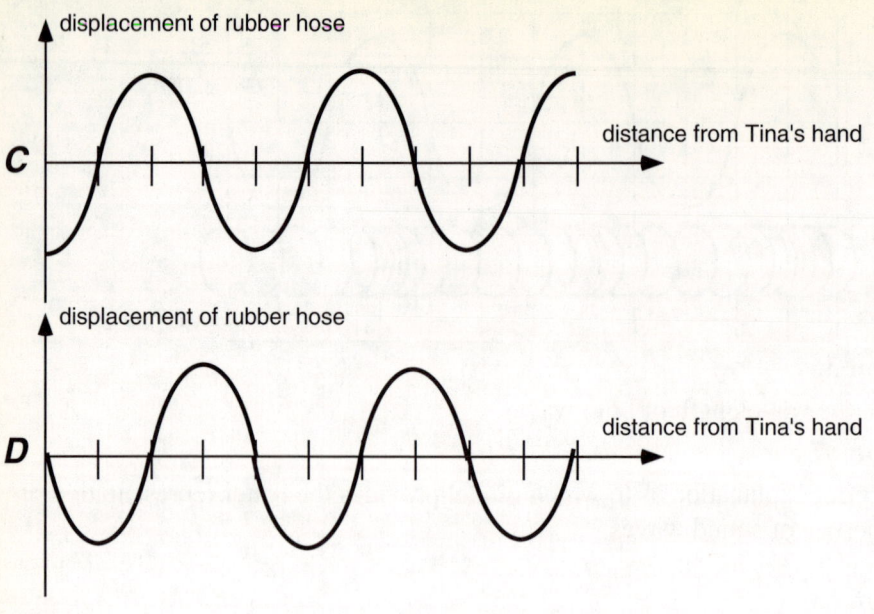

C — displacement of rubber hose vs distance from Tina's hand

D — displacement of rubber hose vs distance from Tina's hand

[1 mark]

A loudspeaker is emitting sound of a fixed intensity which travels equally in all directions. The graph following shows the pressure variation plotted against distance from the speaker at a particular instant of time.

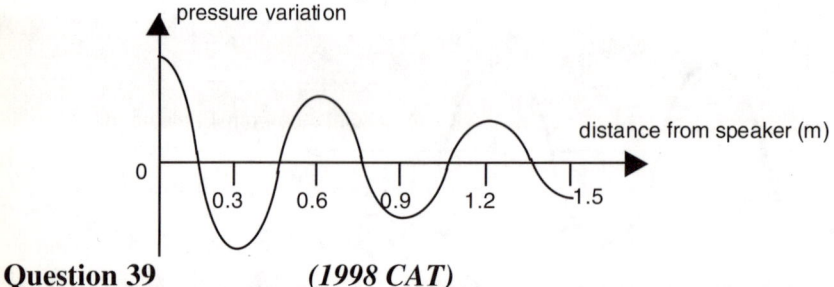

pressure variation vs distance from speaker (m); marks at 0.3, 0.6, 0.9, 1.2, 1.5

Question 39 (1998 CAT)

Explain why the amplitude of the pressure variation decreases as the distance from the speaker increases.

[3 marks]

Question 40 (1998 CAT)

Which of the following best shows the pressure variation at a time that is one-quarter of a cycle later than shown in the graph above?

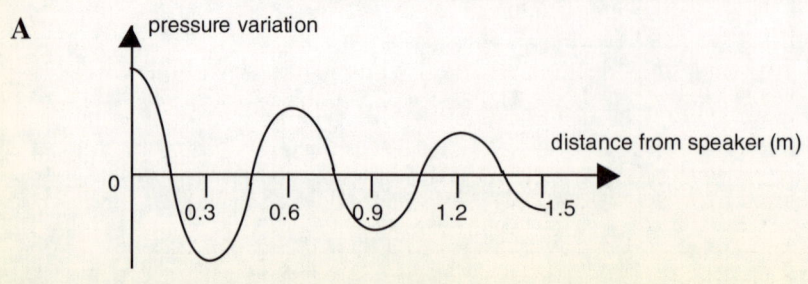

A — pressure variation vs distance from speaker (m); marks at 0.3, 0.6, 0.9, 1.2, 1.5

1 Sound basics

B

pressure variation

0 | 0.3 | 0.6 | 0.9 | 1.2 | 1.5
distance from speaker (m)

C

pressure variation

0 | 0.3 | 0.6 | 0.9 | 1.2 | 1.5
distance from speaker (m)

D

pressure variation

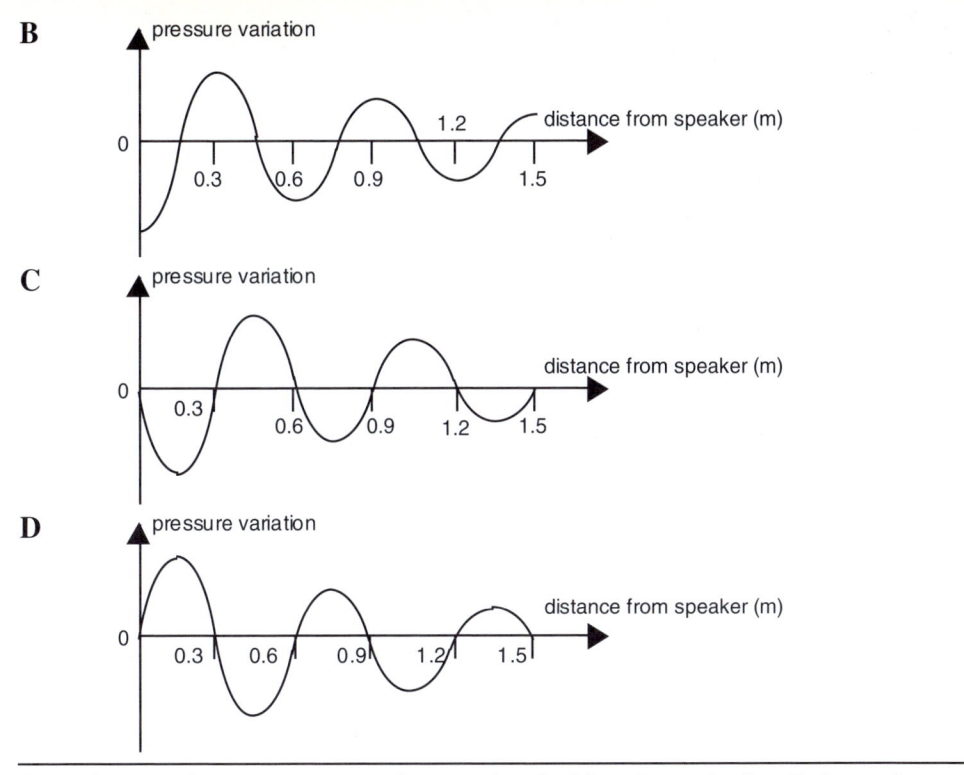

0 | 0.3 | 0.6 | 0.9 | 1.2 | 1.5
distance from speaker (m)

A student produces two types of waves by shaking the end of a slinky spring. The diagram following shows snapshots of the waves.

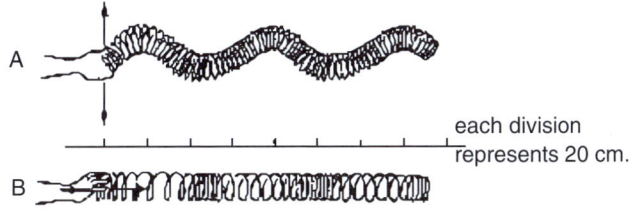

A

each division represents 20 cm.

B

Question 41 *(1998 CAT)*

Estimate the wavelength of travelling wave **A**.

Question 42 *(1998 CAT)*

Which of the waves is a better model for sound waves? Your answer must mention the similarities between the model you choose and sound waves.

[3 marks]

Longitudinal waves on a long spring are used to demonstrate travelling waves. The first part of the spring is shown in the following diagram.

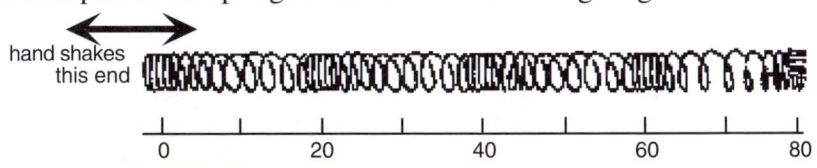

hand shakes this end

0 | 20 | 40 | 60 | 80

1 Sound basics

Question 43 *(1999 CAT)*
Estimate the wavelength of this wave.
[1 mark]

Question 44 *(1999 CAT)*
What is the speed of this wave (in cm s^{-1}) if its frequency is 4.0 Hz?

The cone of a loudspeaker is turned on at t = 0 s, and is driven back and forth so that its *position* as a function of time is as shown in the graph.

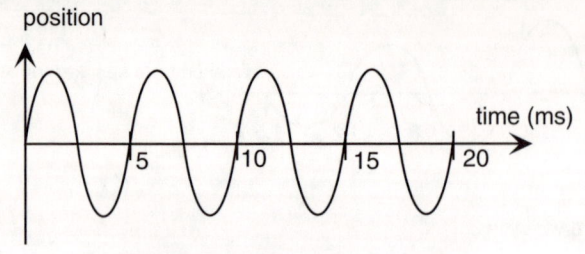

Question 45 *(2001 written examination)*
What is the frequency of oscillation of the speaker cone?

Question 46 *(2001 written examination)*
What is the wavelength of the sound transmitted through the air by the loudspeaker? (Take the speed of sound in air as 340 m s^{-1}.)

A microphone is placed 1.70 m from the loudspeaker and the *pressure* at this point is measured as a function of time.

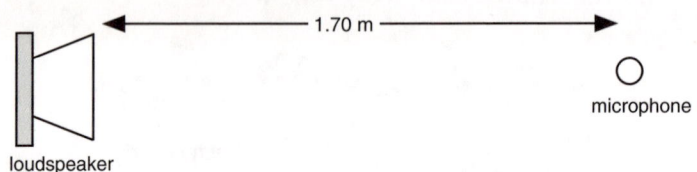

Question 47 *(2001 written examination)*
Which one of the diagrams below best represents the pressure variation at the microphone, as a function of time? The time scale for each starts at t = 0 (when the speaker begins to oscillate). You must justify your answer.

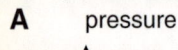

A

14

1 Sound basics

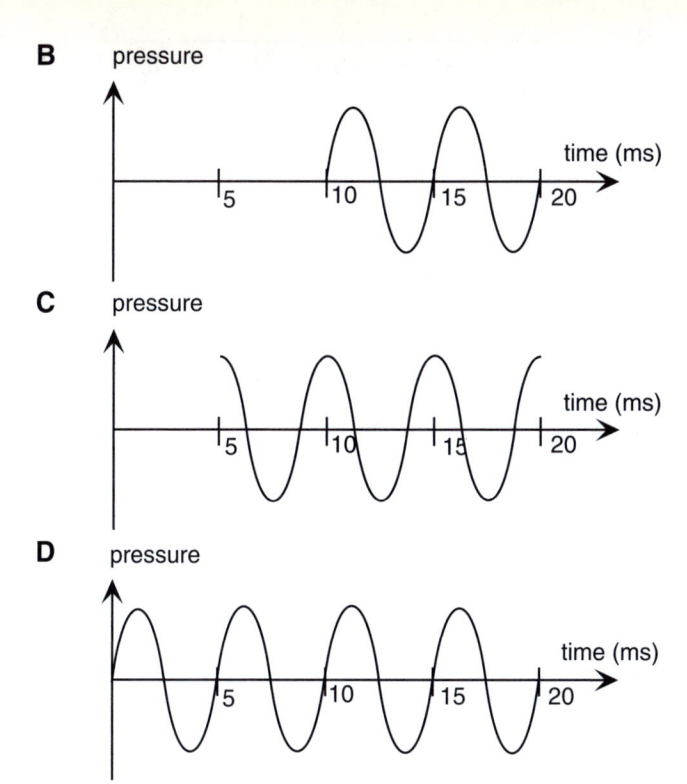

B pressure

C pressure

D pressure

CHAPTER 2   **Sound intensity and level**

Huang has hearing characteristics shown by the curves below. Each curve shows the sound level in decibels (dB) of various frequencies which he hears to have the same loudness. The lower curve is for sounds which he can just hear, and the upper curve is for a level which just causes him pain.

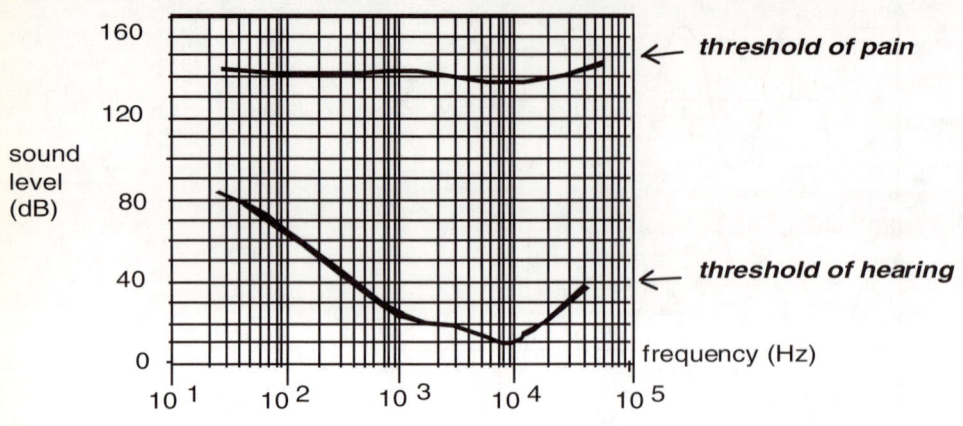

Question 48
There is a difference of about 75 dB between the thresholds of hearing and pain at 100 Hz. Calculate the ratio:

$$\frac{\text{intensity at threshold of pain at 100 Hz}}{\text{intensity at threshold of hearing at 100 Hz}}$$

As part of a test, Huang is exposed to a sound of frequency 100 Hz. The intensity is increased slowly from zero until he can just hear the sound. The frequency is then changed to 500 Hz without changing the output level.

Question 49
By how many dB must the sound intensity level be decreased to the point where Huang can just hear the 500 Hz sound?

Huang is working in a very noisy environment where he is exposed to 1000 Hz sounds of level 140 dB. He is required to wear earmuffs which reduce this sound to a level he can only just hear.

Question 50
By how many dB have the earmuffs reduced the sound level?

Question 51
By what factor have the earmuffs reduced the sound intensity? (Remember that the sound intensity is measured in W m^{-2}.)

Question 52
Amplifiers often have a 'loudness' button which artificially raises the intensity levels at low and high frequencies, compared to mid range frequencies, to compensate for when music is played at low volume. Use Huang's graph above (which is typical) to explain why this is done. [4 marks]

Question 53 (2000 written examination)
The intensity of sound at a microphone a distance of 0.10 m from a singer's mouth is 1.0×10^{-6} W m^{-2}. Her manager is sitting 2.0 m from her mouth. What is the sound intensity at the manager's location? Show your working. (For this problem, assume that sound radiates uniformly in all directions.) [3 marks]

Question 54 (2000 written examination)
The sound system is now turned on, and the intensity at the manager's location increases by a factor of 1000. By how many decibels does the sound increase when the sound system is turned on? [3 marks]

In an experiment to determine hearing sensitivity, a student uses a set of earphones and a signal generator. For a range of frequencies, he determines the sound intensity (measured at the ear) at which the sound intensity just becomes inaudible. The graph below shows the sensitivity of the hearing of the student.

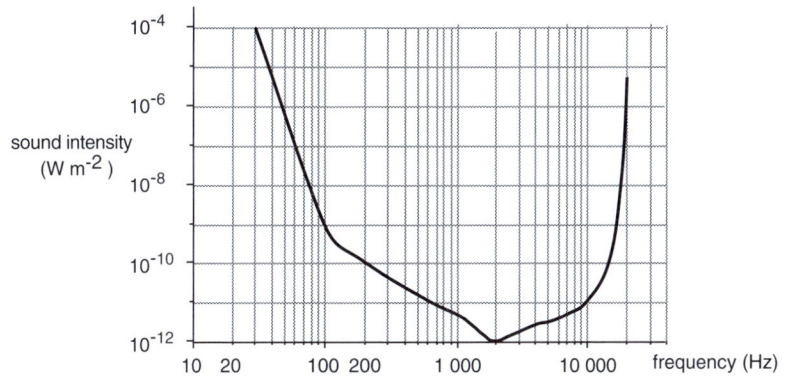

Question 55 (2000 written examination)
At what frequency is the student's hearing most sensitive?

Question 56 (2000 written examination)
What range of frequencies can the student hear if the sound intensity at his ear is 1.0×10^{-11} W m^{-2}?

A teacher is demonstrating to the class that the minimum sound intensity that they can hear depends on the frequency of the sound. A signal generator is used to drive a small loudspeaker. The frequency of the sound is set, and the sound intensity is gradually reduced. Each student records the intensity in W m^{-2} (read from an intensity meter near the loudspeaker) at which the sound becomes audible.

2 Sound intensity and level

The graph following shows minimum sound intensity audible to a person undergoing a hearing test, plotted against frequency.

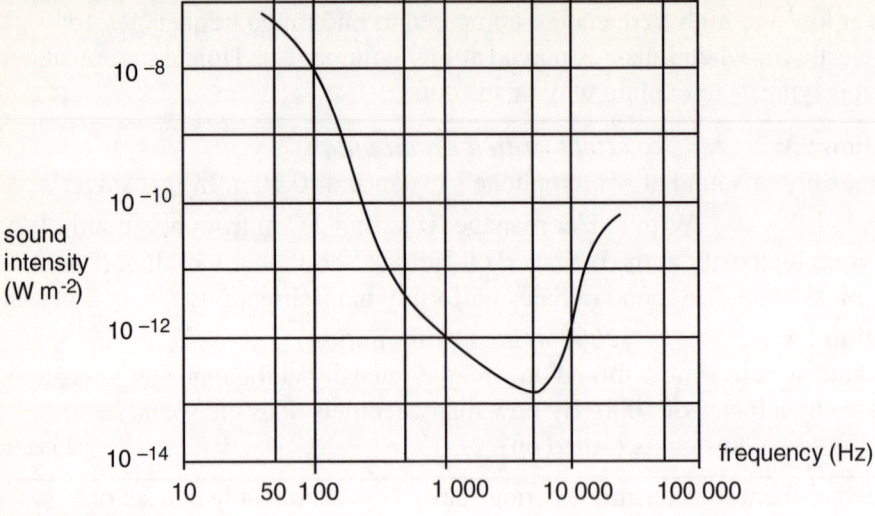

Question 57
At which of the following frequencies is the person's hearing the most sensitive?
A   10 000 Hz
B   1 000 Hz
C   100 Hz

Question 58
What is the value, in units of dB, of the ratio:

$$\frac{\text{intensity at 10 000 Hz that can just be heard}}{\text{intensity at 1000 Hz that can just be heard}} ?$$

Noise from suburban trains can be a problem for nearby residents. The sound intensity 2.0 m from a passing train is 3.5×10^{-3} W m^{-2}.

Question 59
What is the sound intensity level 2.0 m from the passing train?

Question 60
What is the sound intensity due to the train at a distance of 10 m from it? (For this problem, assume sound radiates uniformly in all directions.)

Question 61
If a sound is increased by a factor of 4 in *intensity*, what has happened to the sound *level*?

Question 62
What sound *level* would you associate with a sound of energy intensity of 2×10^{-6} W m^{-2}, if a level of 0 dB is assigned to a sound level intensity of 1×10^{-12} W m^{-2}?

2 Sound intensity and level

Question 63
You are standing 2 m from a band. It's too loud, so you move back so that you are now standing 6 m from the band. What is the (approximate) change in sound level? Show your working.

[4 marks]

The graph below shows the *minimum* sound intensity that a microphone will detect.

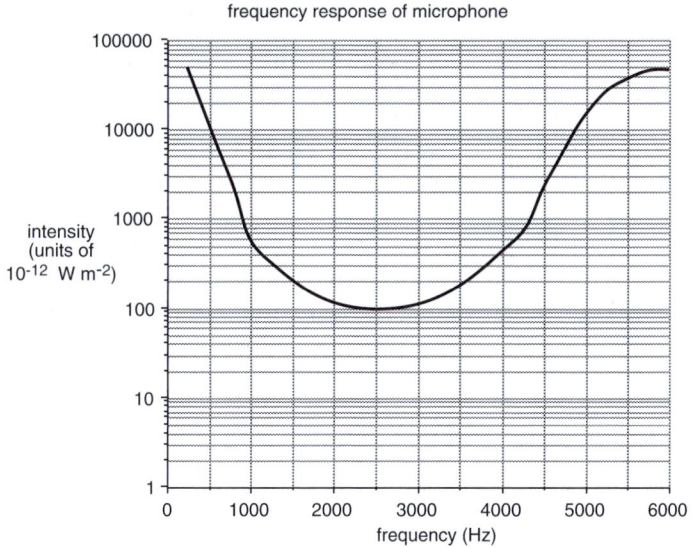

frequency response of microphone

Question 64
What is the ratio:

$$\frac{\text{minimum detectable intensity at 500 Hz}}{\text{minimum detectable intensity at 2500 Hz}} ?$$

Give your answer in dB.

Question 65
If 0 dB is taken as equivalent to an intensity of 10^{-12} W m^{-2}, what is the value of the minimum sound level detectable by the microphone at a frequency of 1000 Hz?

Question 66
A sound system resonates at a frequency of 150 Hz, causing this frequency to be reproduced with an intensity *five* times as great as neighbouring frequencies. Express this difference in intensity in dB.

Read the following passage.

A solution to the frequency response problem in loudspeakers can be to use several speakers to span the audible range. A woofer is used for 20–500 Hz; a mid-range speaker for 500–5 kHz, and a tweeter for the frequencies from 5–20 kHz. The crossover frequencies are typical values; they vary from system to

2 Sound intensity and level

system. The graphs following summarise the situation. A crossover network sends the output from the amplifier to the speakers. Each speaker only receives the frequency range to which they respond.

frequency response of low-frequency 'woofer', midrange speaker and high frequency 'tweeter'

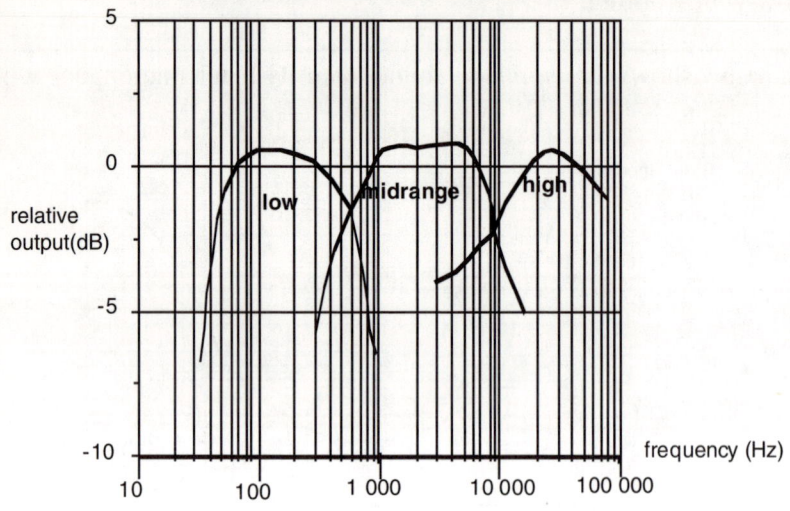

Question 67
What is the relative output from the 'tweeter' at 10 000 Hz?

Question 68
At what frequency are the relative outputs of the 'woofer' and the mid-range speakers equal?

Question 69
When the woofer output level drops 6 dB below its maximum, it is producing noticeably less sound intensity. What is the value of the ratio:

$$\frac{\text{sound intensity at maximum value}}{\text{sound intensity 6 dB below maximum value}}?$$

Question 70
The human ear can just detect a change in sound level of 3 dB. Estimate, from the graph, the frequency range of the mid-range speaker where the relative output of the speaker does not detectably change.

Jimi stands 10 m from a music group. Tina is 35 m from the group. Tina carries a sound intensity meter with her. It reads 5×10^{-4} W m^{-2}. Assume that the sound from the band spreads out uniformly.

Question 71
Estimate the sound intensity in W m^{-2} at the spot where Jimi is standing.

Question 72
Estimate the total acoustical power given out by the music group.

Question 73

If an intensity of 1×10^{-12} W m^{-2} is taken as 0 dB, what is the sound intensity level at the point where Tina is standing?

Question 74

Arrange the following in order of *increasing* sound level intensity.

A	The noise which Tina hears.
B	A fairly quiet classroom.
C	1 m from a jack hammer (mending the roads).
D	Rustling leaves in a light breeze.

[1 mark]

Question 75

By how many dB is the noise Jimi hears louder than the noise Tina hears?

The graph following shows the approximate range of human hearing.

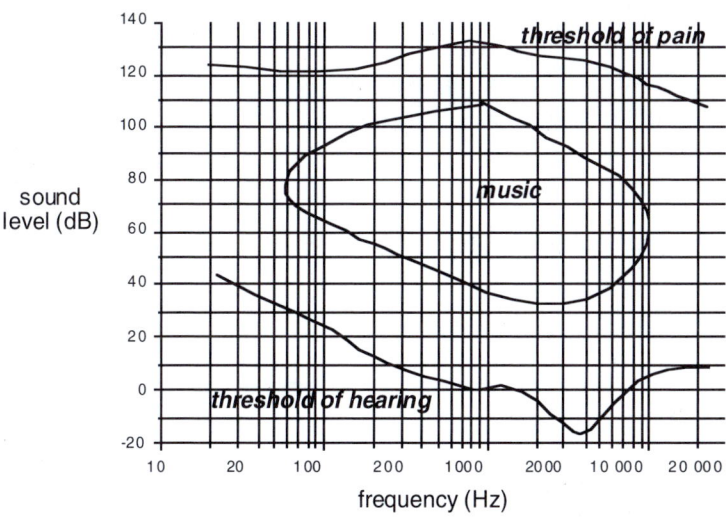

Question 76

Estimate from the graph the threshold of hearing (in dB) at 400 Hz.

Question 77

If 0 dB is defined as the level at an intensity of 1×10^{-12} W m^{-2}, what is the sound intensity of a 400 Hz sound at the threshold of hearing?

Question 78

At a frequency of 2000 Hz, what is the difference (in dB) between the loudest and softest *music* as shown on the graph?

Question 79

If the sound intensity of the loudest 2000 Hz music is labelled I_L, and the softest 2000 Hz music is labelled I_S calculate the ratio: I_L/I_S

2 Sound intensity and level

Question 80
A car has an '8 speaker sound system'. If each of the speakers contributes an equal amount of power, by how many dB is the sound level from eight speakers greater than that from one speaker?

Question 81
Kurt stands 5 m from a 100 W sound system, operating at full volume. Calculate the sound intensity in W m^{-2} where Kurt is standing. (Assume that the sound spreads out evenly in all directions.)

Question 82
Decide whether the sound level intensity Kurt hears is louder or softer than that of a reasonably noisy classroom. Justify with a calculation.

Question 83
If the sound from the sound system was *not* spread over all directions, but confined only to *one quarter* of the available directions, by how much would the sound level intensity Kurt hears increase?

In tests for material for earmuffs, sound is directed at a sheet of material, as shown. A sound intensity meter at A registers a reading of 5.0×10^{-8} W m^{-2}; at B the reading is 7.0×10^{-9} W m^{-2}.

Question 84
What is the difference in sound level between A and B?

Question 85
How much energy per second has been absorbed per square metre in the earmuff material?

Roofing material is tested acoustically by sending pulses of sound waves from a sound generator. A model of the process is sketched at the right. An intensity of I_P is incident on the roofing material. Of this, I_A is reflected, and I_B transmitted.

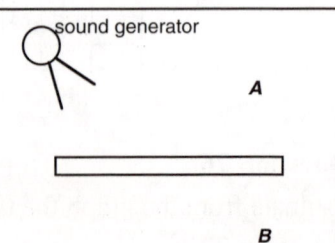

Question 86
Write an expression for the intensity absorbed by the material.

Question 87
Which of the following best describes the *absorption* of energy within the roofing material under test?

A   The regular vibration of sound waves outside the material is simply transferred to regular vibrations within the material.

B   The molecules of the material do not vibrate as easily as the air molecules, so the energy is simply lost.

C	The regular vibration of molecules associated with sound waves is transformed to random vibrations within the material.
D	All the energy inside the material is eventually re-radiated as sound energy outside the material.

Question 88

A new material is claimed to 'absorb 90% of the sound energy that falls upon it'. How many dB is this equivalent to?

Question 89

Another company advertises that its sound insulation material 'reduces sound levels by a massive 20 dB'. What percentage of incident energy is absorbed if their claim is correct?

Question 90

A loudspeaker radiates sound more or less equally in all directions at a football stadium. People sitting near the loudspeaker hear it as very loud, yet those sitting on the other side of the arena hear it as fairly quiet.

Select the *best* reason for this from the list below.

A	Sound energy is absorbed significantly by air.
B	Interference effects account for the difference.
C	Sound energy is spread over a larger area at greater distance from the speaker.
D	Diffraction of different frequencies is the main reason for this effect.

Noise from passing traffic on a freeway is a problem for nearby residents. The sound intensity, measured at a distance of 1.0 m from a passing car is 1.0×10^{-3} W m^{-2}.

Question 91   (1996 CAT)

What is the sound intensity level 1.0 m from the passing car, expressed in units of dB?

Question 92   (1996 CAT)

What is the sound intensity, due to the car, at a house 100 m from the car (in units of W m^{-2})? Assume that the source of the sound is small, and that the sound radiates equally in all directions.

The graph following represents the sensitivity curve for hearing for an average person; that is, the lowest intensity sound that can be heard at a given frequency.

2 Sound intensity and level

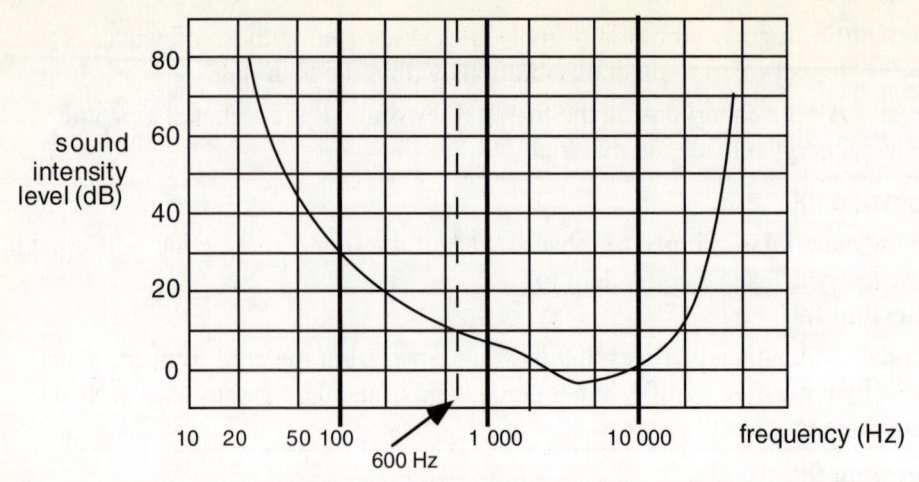

To test the sensitivity of hearing at various frequencies, a student with normal hearing sits near a small speaker, connected to a frequency generator tuned to 200 Hz. The intensity is increased from 0 until the student can just hear the sound. The generator is now tuned to 600 Hz without changing the output level (see the dashed line above).

Question 93   *(1997 CAT)*

By how many dB must the sound intensity level be decreased so that the student can just hear the sound?

A band is playing in the school hall. A student 15 m from the band hears the music clearly. In the afternoon the band plays the same music outdoors. The student is again seated 15 m from the band. She notices that the band sounds softer, and that the sound lacks low and high frequencies.

Question 94   *(1997 CAT)*

Explain why moving the performance outdoors has reduced the sound intensity at the listener's location.

Question 95   *(1997 CAT)*

With reference to the graph above, explain why the high and low frequencies are not heard by the student when the band plays outside.

Question 96   *(1999 CAT)*

Which of the following would produce at your ears a sound intensity level of about 100 dB?

- A   a police car passing near you in the street with its siren on
- B   rain falling on your umbrella
- C   40 m from a jet at take off
- D   normal conversation

2 Sound intensity and level

Question 97   *(1999 CAT)*
The 4000 Hz siren at the CFA headquarters is mounted on a high stand, as shown. On a windless day, the intensity at the point Q (which is 20 m from the siren), is 0.0040 W m^{-2}. What is the sound intensity level at this point Q? (Ignore reflections from the ground.)

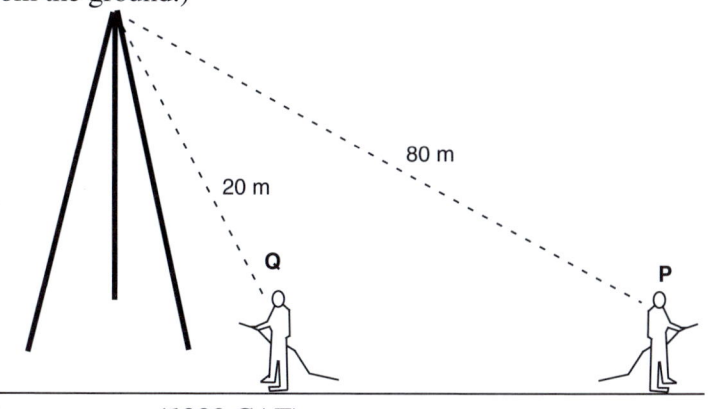

Question 98   *(1999 CAT)*
What is the intensity of the sound at the point P, 80 m from the siren?

Question 99   *(1999 CAT)*
Some sound is in fact reflected from the ground. If this is taken into account, would your answer be greater, less or the same?

[3 marks]

The graph following shows the sensitivity of the ear of an average person as a function of frequency. Sound intensity levels below the threshold of hearing cannot be heard, whilst those above can be heard.

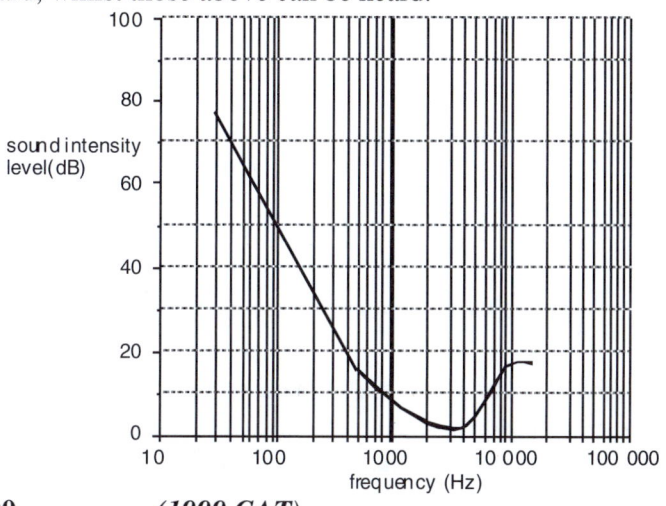

Question 100   *(1999 CAT)*
Explain, making use of the graph, why the frequency of the siren was chosen to be around 4000 Hz.

2 Sound intensity and level

In a demonstration of the perception of loudness, a teacher sets up a loudspeaker on a stand at the centre of the school oval. The loudspeaker emits sound equally in all directions with a wavelength of 1.0 m. Ignore reflections from the ground.

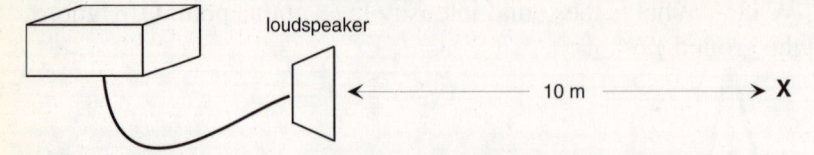

Question 101 *(2001 written examination)*

Xena, a student, stands at the point **X**, 10 m from the loudspeaker, and measures the intensity of the sound to be 9.0×10^{-8} W m^{-2}. she then moves to a place further away from the loudspeaker, and measures the intensity of the sound to be 2.25×10^{-8} W m^{-2}.

How far is Xena from the loudspeaker at this new position?

Question 102 *(2001 written examination)*

By how many decibels has the intensity level of the sound changed between the two readings?

Mel and Jill attended an orchestral concert in a large cathedral with stone walls and ceiling. They noticed that after the orchestra stopped playing they could still hear the sound for some time. They also observed that the low-frequency and high-frequency sound persisted for a *shorter* time than sound in the middle frequency range.

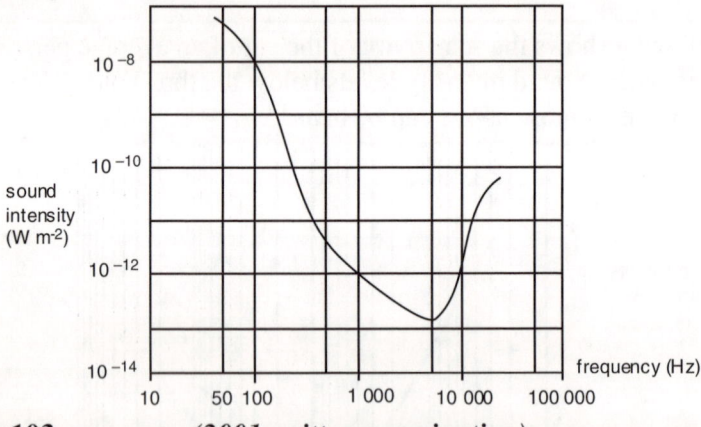

Question 103 *(2001 written examination)*

Explain why:
• the sound persisted after the orchestra stopped playing
• the high-frequency and low-frequency sounds were heard for a shorter time.
You may need to refer to the graph above, which shows the minimum detectable sound as a function of frequency of an average human ear. [4 marks]

Question 104 *(2001 written examination)*

What is the frequency of the sound they could hear for the longest after the orchestra stopped?

CHAPTER 3   Standing waves and resonance

Jannie is standing between two loudspeakers at a sporting ground. They are being tested with a note of a single frequency of 340 Hz. The loudspeakers emit sound reasonably well in all directions.

Question 105
When she stands midway between the speakers the sound is quite loud, but as she moves away from this point towards either speaker the sound gets softer. Explain why this happens.

[4 marks]

Question 106
How far must she move from the centre point towards either speaker before the sound becomes loud again? (Speed of sound = 340 m s^{-1}.)

Parasaurolophous dinosaurs made sounds from hollow horns on their heads. These can be modelled as 2.80 m pipes open at both ends.

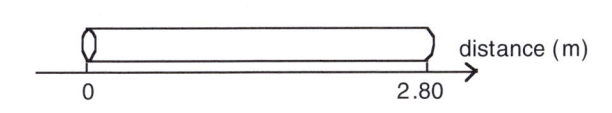

Parasaurolophous dinosaur

The graph following shows pressure variation of the air inside this pipe when a sound of fundamental frequency f_0 is being emitted. The two lines represent the maximum and minimum values of the pressure variation.

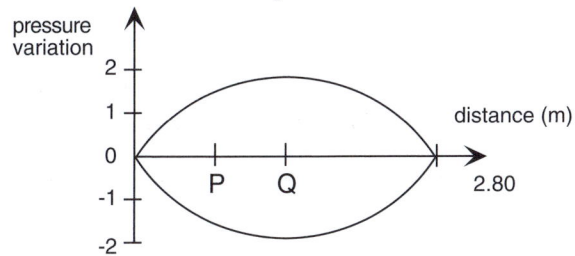

3 Standing waves and resonance

Question 107 *(1999 CAT)*

What is the wavelength of the sound produced under this condition?

A pedestrian tunnel near a busy road sometimes resonates at a low frequency when there is enough background noise. A particular tunnel is 3.1 m long, and the resonant frequency observed is close to 55 Hz.

Question 108

Explain why this frequency is likely to be produced by a tunnel of this length if it is modelled as a tube open at both ends, and the speed of sound is taken as 340 m s^{-1}. Include calculations in your answer.

[5 marks]

Question 109

The traffic noise contains many different frequencies, but the resonating frequency is quite specific. Why does the tunnel not resonate to most of these other frequencies?

[3 marks]

An empty bottle resonates strongly to a frequency of 512 Hz. It also resonates (but less strongly) to frequencies of 1536 and 2560 Hz. However it does not resonate to frequencies in between these values.

Question 110

Suggest an explanation for these observations. You should model the empty bottle by a pipe which is either open at both ends or closed at one end.

[4 marks]

Question 111

If the empty bottle were to be partly filled with water (assuming it does not leak), what do you predict would happen to the resonant frequencies observed previously?

[3 marks]

The vocal tract can be modelled by a tube of length L that is open at one end and closed at the other, as shown on the diagram below. The fundamental frequency of the tube is 500 Hz.

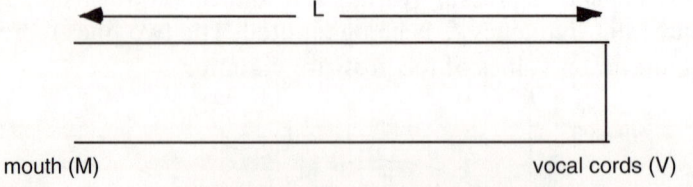

mouth (M)   vocal cords (V)

Question 112 *(2000 written examination)*

Calculate the length L of the tube. The speed of sound is 340 m s^{-1}.

Question 113   (2000 written examination)
Which *one or more* of the following frequencies would also resonate?
A   250 Hz   B   1 500 Hz   C   1 000 Hz
D   2 000 Hz   E   2 500 Hz

Question 114   (2000 written examination)
A singer emits a pure sound of frequency 500 Hz. Which one of the following diagrams best shows the *maximum* pressure variation (ΔP_{max}), above and below normal atmospheric pressure (P_0), along the tube which models the vocal tract? (The letters M and V indicate the location of the mouth and vocal chords in the diagram above.)

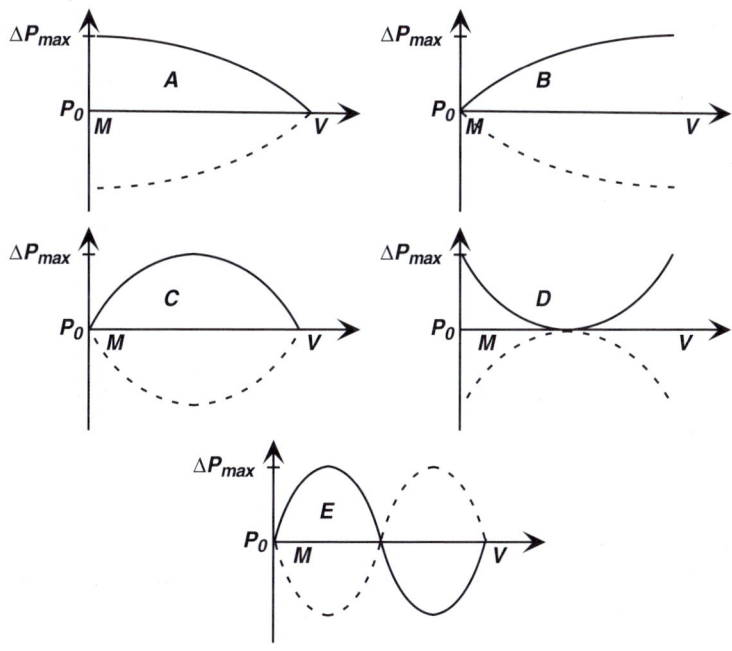

Question 115
A clarinet can be modelled as a pipe with one open end and one closed end. It is playing a note of fundamental 440 Hz. Explain, using diagrams, why it *cannot* produce a harmonic of frequency 880 Hz. The speed of sound is 335 m s^{-1}.
[4 marks]

Question 116
A flute plays the same note as the clarinet in the previous question. It can, however, produce a harmonic of frequency 880 Hz. Explain, using a diagram.
[4 marks]

3 Standing waves and resonance

A flute can be modelled by a column of air, open at both ends, as shown.

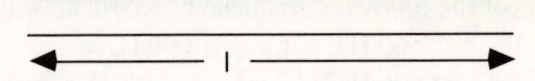

A new 'bass' flute is designed to play very low notes. The lowest note of its range will have a f = 65 Hz. Take the speed of sound to be 335 m s^{-1}.

Question 117
Which of the following is the best estimate of its length? Justify your answer with calculations.

A   5.0 m   B   3.5 m   C   2.5 m   D   1.5 m

[4 marks]

Question 118
If the fundamental of the flute is 65 Hz, what are the next *two* harmonics (sometimes called overtones) that the flute should be able to play?

A large pipe is set up in a museum. It is open at both ends, as shown below. It is large enough for people to walk along inside it. A large loudspeaker is set up at one end. The diagram is not drawn to scale. Jin is walking along the pipe.

During a demonstration, one frequency resonates strongly in the pipe.

Question 119
Explain what is meant by 'resonates'.

[4 marks]

As Jin walks along the pipe, he encounters spots where the resonating sound is very soft (nodes). They are 2.5 m apart, and there are a total of 3 of them inside the entire length of the pipe. The two ends of the pipe are also soft spots.

Question 120
Explain why these nodes (the very soft spots) occur. You should refer to the principle of superposition in your explanation.

[4 marks]

Question 121
What is the wavelength of the resonating sound?

Question 122
As he walks along the pipe he also encounters spots where the sound is at a maximum. What is the spacing between these loud spots (antinodes)?

Question 123
Which of the following is the best estimate for the length of the pipe?

A   6 m   B   7 m   C   10 m   D   15 m

3 Standing waves and resonance

Question 124
If the speed of sound is taken as 350 m s⁻¹, what is the frequency of the resonating sound?

Question 125
If the answer to Question 121 is denoted λ_1, which *one or more* of the following frequencies would also be likely to resonate in the pipe?

A $3\lambda_1$ B $3/2\,\lambda_1$

C $3/4\,\lambda_1$ D $2\lambda_1$

Question 126
In a later demonstration, the end of the pipe furthest from the loudspeaker is closed. Which of the following best describes what this would do to the *number* of resonances below 90 Hz?

A   It would increase the number of resonances.
B   It would decrease the number of resonances.
C   There would be the same number of resonances.

Give clear reasons, including calculations, for your choice of answer.   [4 marks]

Question 127
Syd is given the choice of two pipes for a simple musical instrument. One is open at both ends, the other closed at one end. Which one will produce the *lowest* note? Justify your answer. A diagram is recommended.   [5 marks]

Question 128
Which *one or more* of the following formulae best describes the sequence of harmonics for a pipe of length l open at both ends? (NOTE: n = 1, 2, 3, 4, 5,...). c stands for the speed of sound.

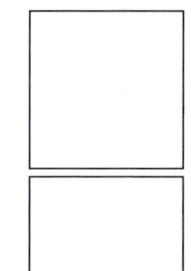

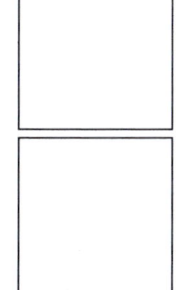

A   [ ]   B   [ ]

C   [ ]   D   [ ]

Question 129
Which *one or more* of the above formulae best describes the sequence of harmonics for a pipe of length l closed at one end and open at the other?

Students are exploring why a bugle can be used to produce a range of notes, even if it is of fixed length. They model the bugle and player by using a length of pipe as shown, with a sound source placed at S as shown. They expect that this system will act as a pipe open at one end.

S.

They find that, as they increase the frequency of the source S, the pipe emits a loud sound for frequencies corresponding to wavelengths of 1.2 m, 0.4 m and 0.24 m.

Question 130
What is the length of the pipe? Show your working.

[3 marks]

With a bugle the lips of the player will vibrate with a *range* of frequencies.

Question 131
Explain why the sound emitted by a bugle will always consist of more than one frequency, and why the instrument can be used to play a simple sequence of notes.

Question 132
A violin string is tuned to 440 Hz. It is attached firmly at both of its ends. Explain, using diagrams, why this means that it can produce harmonics of frequency 880 Hz and 1320 Hz.

[4 marks]

A trombone player uses vibrating lips to generate sound, and then tunes the instrument to the desired resonant frequency by sliding one section over the other to adjust the length of the tube. The trombone can be modelled as a tube of variable length as shown following.

To investigate this model, students put a single frequency sound source at S. They vary the length of the tube while observing the intensity of sound at the other end. The intensity goes through a series of maxima and minima as the length changes. The students observe that successive maxima occur each time the length changes by 0.115 m.

Question 133   (1997 CAT)
What is the wavelength of the sound?

Question 134   (1997 CAT)
In a real trombone, the sound from the lips of the player is of relatively low intensity and consists of a range of frequencies. Explain how a trombone is used to produce *loud* sounds of different single frequencies needed to play a melody.

[4 marks]

3 Standing waves and resonance

An organ pipe making a sound of frequency 250 Hz is very close in shape to a cylinder open at both ends. The sketch following shows the approximate distribution of molecules in the organ pipe under test at one time. There is a *standing wave* formed in the organ pipe. The speed of sound is 345 m s^{-1}.

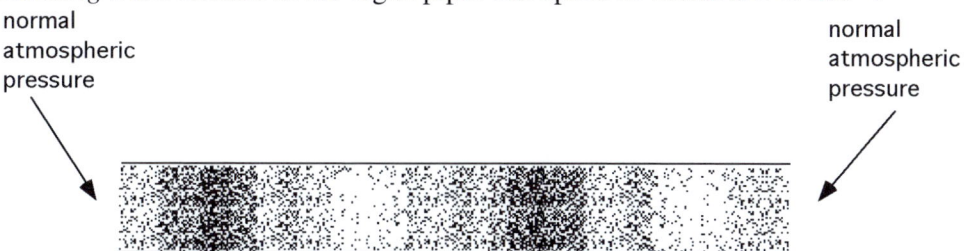

normal atmospheric pressure

normal atmospheric pressure

Question 135
Sketch a graph showing how the pressure depends on the distance along the organ pipe at the time the sketch above was made.

Question 136
Calculate the length of the organ pipe.

Jack uses a tube closed at one end to model a wind instrument. By changing the frequency of a small loudspeaker very close to the open he creates resonances at several different frequencies.

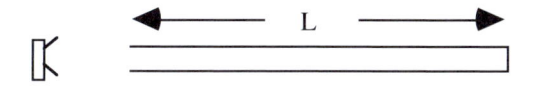

L

(not to scale)

The four lowest frequency resonances have wavelengths of 0.84 m, 0.28 m, 0.17 m. The speed of sound in air is 340 m s^{-1}.

Question 137
What is the lowest frequency at which resonance occurs for this tube?

Question 138
What is the length of the pipe?

Question 139
If the students were to record another resonance, what wavelength reading would you predict?

Question 140
If the same tube was used by students studying physics at high altitude in Nepal, where the speed of sound is 10% greater than at sea level, would the wavelengths at which resonance occurs be the same, shorter or longer than at sea level? Explain your answer.

[3 marks]

Question 141
A flute soloist travels to Nepal to accompany a local orchestra. What are the practical implications of the greater speed of sound at high altitude for the pitch of this flute compared to a piano, for instance?

[4 marks]

A didgeridoo, an instrument used first by Australian Aboriginal people, may be modelled by a cylindrical pipe, open at both ends. A 1.25 m long didgeridoo is modelled below, as shown.

The graph below shows the pressure variation of the air inside the pipe when a sound of fundamental frequency f_0 is being emitted. The two lines represent the maximum and minimum values of the pressure variation.

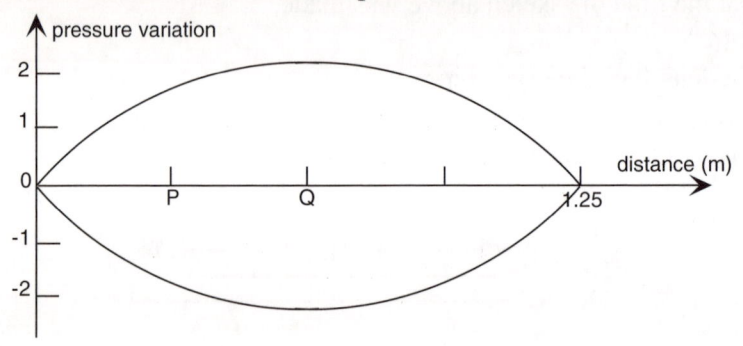

Question 142   (1999 CAT)
What is the wavelength of the sound produced under these conditions?

Question 143   (1999 CAT)
What is the frequency f_0? Take the speed of sound in air to be 340 ms^{-1}.

The graph following shows the pressure variation as a function of time at the middle of the pipe (point Q in the previous graph).

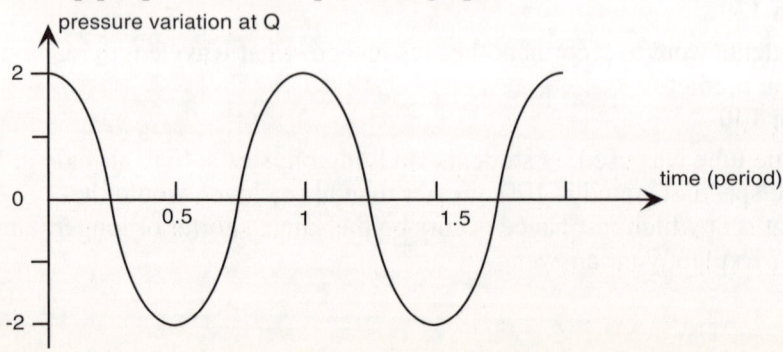

Question 144   (1999 CAT)

Which of the following graphs best shows the pressure variation at the point P as a function of time?

A. pressure variation at Q

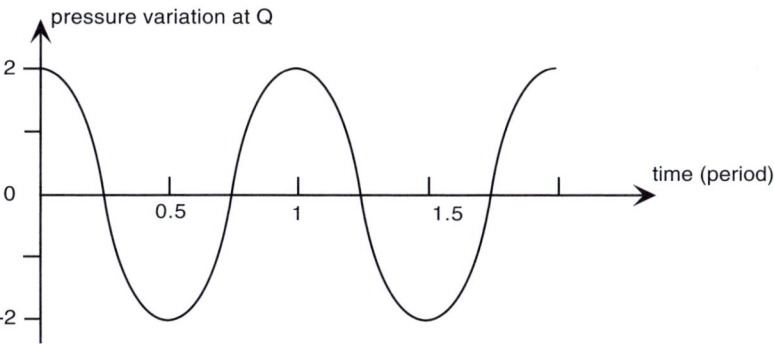

B. pressure variation at Q

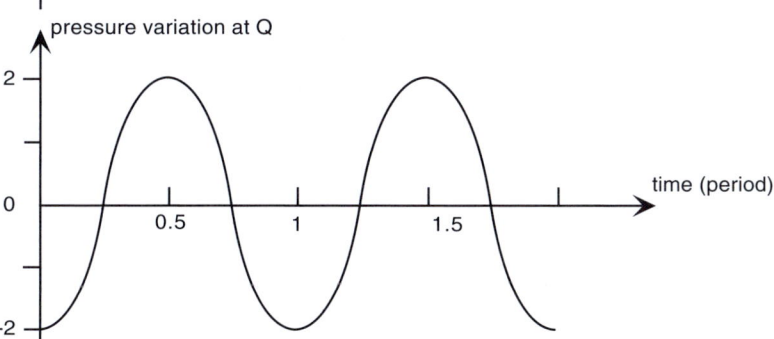

C. pressure variation at Q

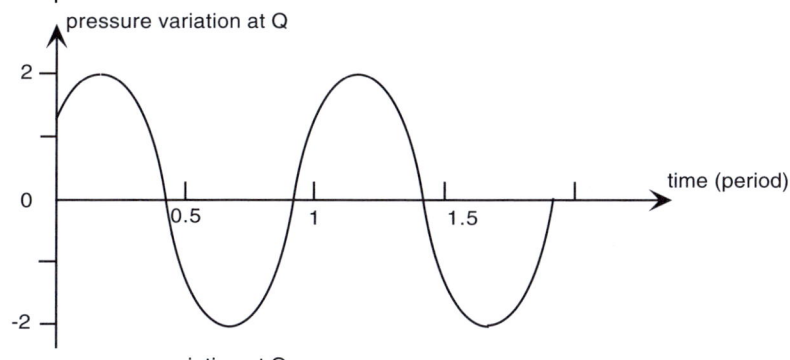

D. pressure variation at Q

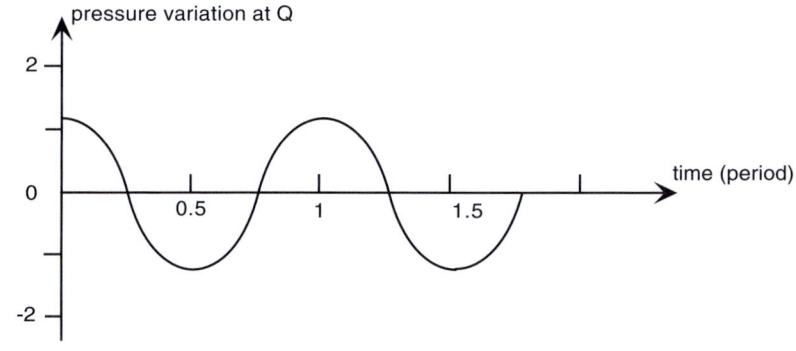

3 Standing waves and resonance

Sound of only a few frequencies can be generated by the didgeridoo. The pressure variation for the lowest possible frequency f_0 is shown in the graph in Question 142.

Question 145   *(1999 CAT)*

Which of the following graphs best shows the maximum and minimum values of the pressure variation along the tube for the next higher frequency f_1 above f_0 ?

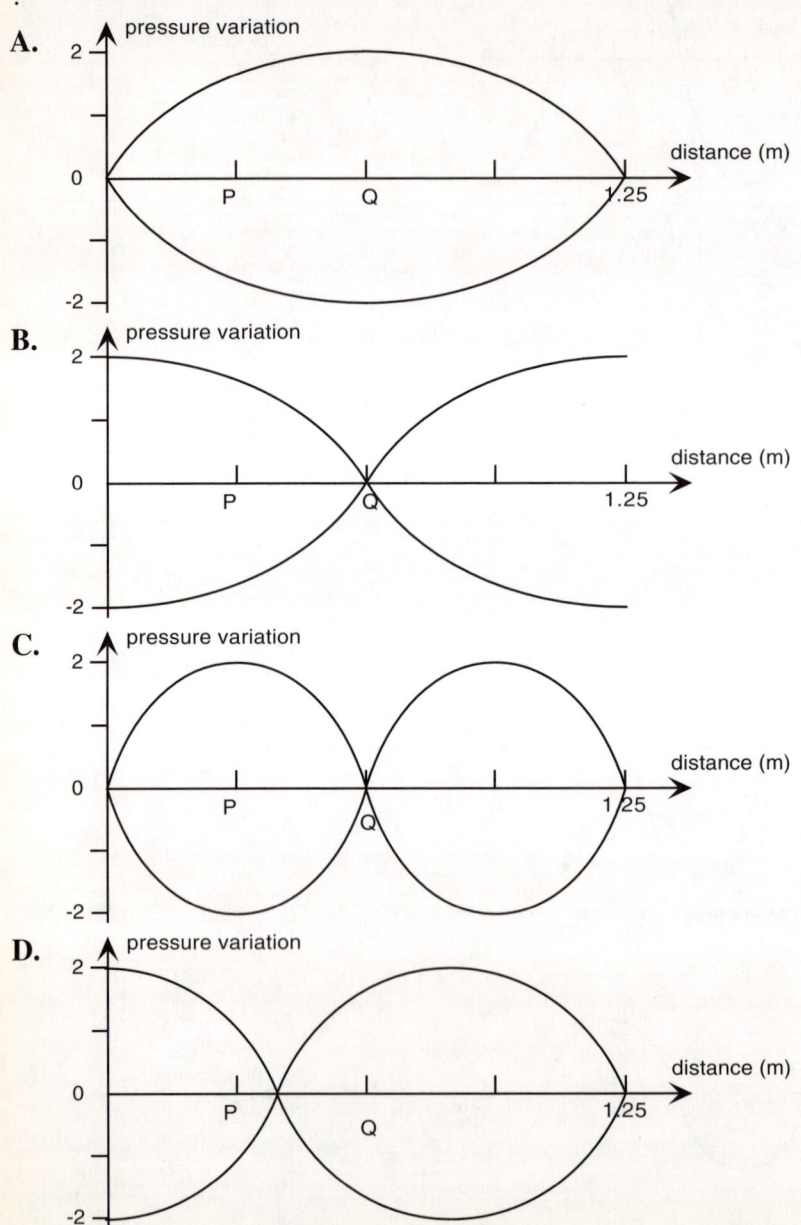

A.

B.

C.

D.

3 Standing waves and resonance

Students use a narrow tube of length 0.432 m open at one end to model a flute. By varying the frequency of sound emitted from a small loudspeaker placed near one end, as shown below, they observe several resonances.

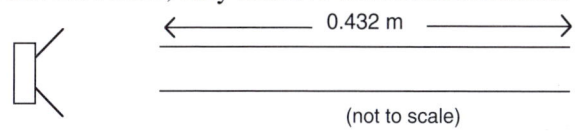

(not to scale)

The wavelengths of the sound at which the resonances with the three lowest frequencies occur are 0.864 m, 0.432 m and 0.288 m. The speed of sound in air is 340 m s^{-1}.

Question 146   *(1998 CAT)*

What is the lowest frequency at which resonance is observed by the students? Show your working.

[3 marks]

The students then fill the tube with helium gas, which has a speed of sound of 1000 m s^{-1}.

Question 147   *(1998 CAT)*

The students find the longest wavelength of sound which produces resonance in the tube filled with helium. Is this wavelength the same, longer, or shorter than the longest wavelength when the tube was filled with air? Explain your answer.

[3 marks]

Standing waves are set up in a narrow glass tube, as shown on the following diagram. An audio signal generator and a small speaker are used to set up the standing waves in the tube. The frequency is adjusted to set up the resonances. The tube is filled with fine dust so that when a resonance is formed the dust indicates the positions of the pressure nodes and antinodes. Although the entire tube is visible, shields prevent seeing whether the ends of the tube are open or closed.

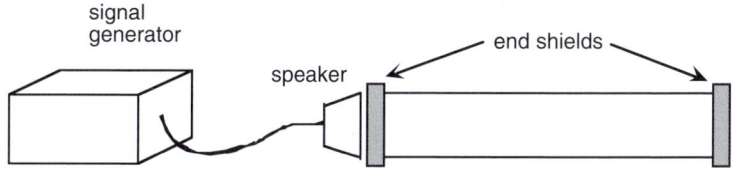

Question 148   *(2001 written examination)*

At a particular frequency, there are 5 nodes and 5 antinodes in the tube. How many open ends does the tube have? Include a diagram to justify your answer.

Question 149   *(2001 written examination)*

What is the length of the tube? Take the speed of sound in air to be 340 m s^{-1}.

[4 marks]

CHAPTER 4   **Diffraction and interference**

A demonstration of interference is set up on a school playing field. Two loudspeakers, S_1 and S_2, are connected to the same source, and produce sound of single frequency, in phase. A line XY is marked on the field, 16 m from the line S_1S_2. A line BC is perpendicular to the line joining the speakers, and point B is midway between S_1 and S_2. The situation is shown on the right.

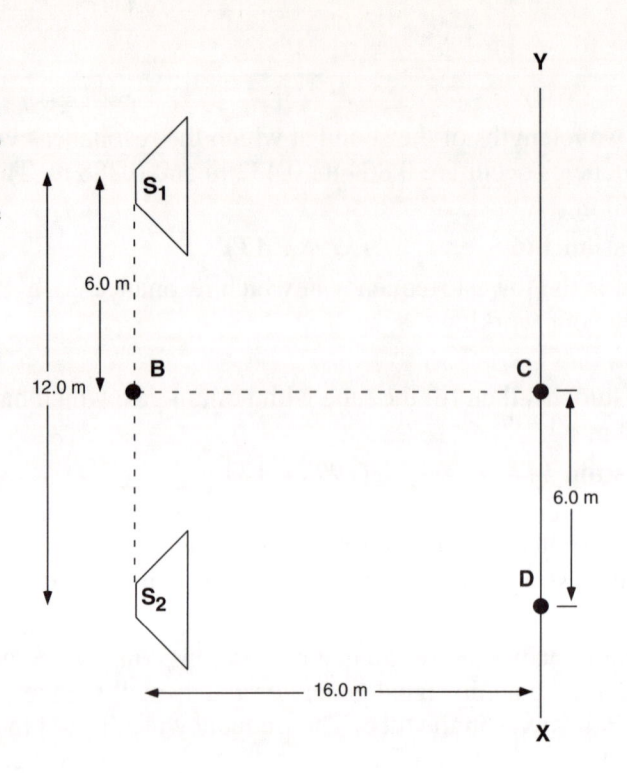

Question 150   *(2000 written examination)*

A student walks from point C towards point B, measuring the sound intensity as she goes. which *one* of the statements below best describes the sound intensity as she walks for point C to point B?

A   The sound intensity remains constant.
B   The sound intensity gets louder as she gets closer to point B.
C   The sound intensity varies between loud and soft, and the variation increases as she gets closer to point B.
D   The sound intensity varies between loud and soft, but the variation is the same as she gets closer to point B.

Question 151   *(2000 written examination)*

The student now walks from point C towards X. She detects maxima and minima in the sound intensity. At the point D she detects the *second minimum* after leaving point C. What is the wavelength of the sound produced by the loudpeakers? Show your working.

[4 marks]

4 Diffraction and interference

Max and Michelle are buying loudspeakers for their hi-fi. They are choosing between two models, a 35-cm diameter speaker (P) and a 5-cm one (Q). Both speakers operate equally well over the complete audible frequency range. Max is standing in front of the speakers, and Michelle is the same distance away, but to one side, as shown in the diagram below.

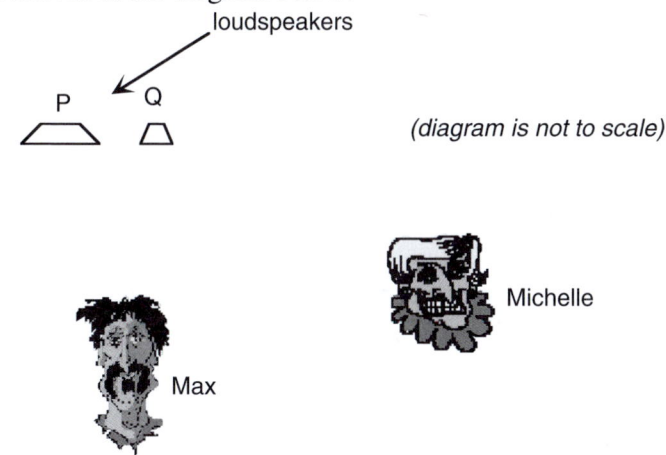

(diagram is not to scale)

As a test of the speakers they play sounds of 10 000 Hz and 200 Hz, and compare the intensity that they each hear.

Question 152   *(2000 written examination)*

The wavelength of sound of frequency 200 Hz is 1.65 m. What is the wavelength of sound of frequency 10 000 Hz?

Question 153   *(2000 written examination)*

Max comments that the intensity of the sound of both frequencies seems the same from both speakers. Michelle says that for the larger speaker, the 10 0000 Hz sound is significantly softer than the low frequency sound. Explain these observations. Include relevant calculations and/or diagrams.   [4 marks]

Two loudspeakers are attached to a school building. They are both emitting the same frequency noise (a 640 Hz tone). Fred has managed to leave class a little early, and is walking along the line *ABCDE*. He notices that the intensity of the sound varies. It is loud at *A*, soft at *B*, loud at *C*, soft at *D*, and loud at *E*. The points *A, B, C, D* and *E* are nearly equally spaced.

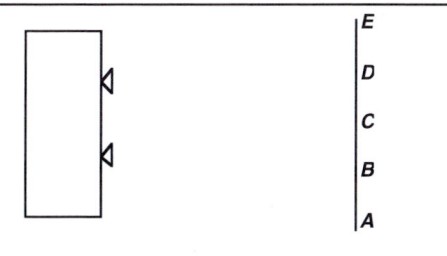

Question 154

Explain why the intensity varies as described above. Use diagrams.   [4 marks]

Question 155

A is 16.0 m from the nearest speaker. The speed of sound is 320 m s^{-1}. How far is the point *A* from the other loudspeaker? Show your working.   [4 marks]

4 Diffraction and interference

Students are at a concert with their parents. The parents find the sound too loud and wait outside. A plan of the concert hall (which has thick, sound-proof walls) is shown. As they walk out and reach point X, they find that the sound level drops. At Y, beyond the direct line to the stage, the sound level is even lower.

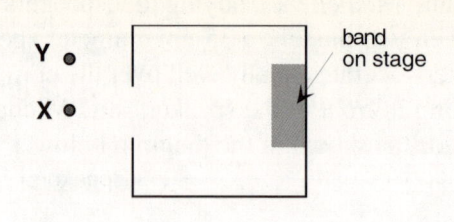

Question 156
They also find that the intensity of the high-frequency sound is relatively much weaker than the low-frequency sound. Explain why this is so. [4 marks]

A noisy orchestra is playing in a hall as shown to the right.

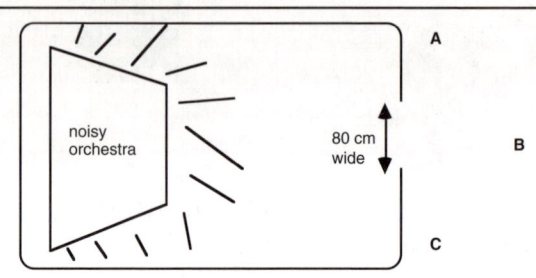

Question 157
The door at the end of the hall is 80 cm wide. The speed of sound can be taken as 320 m s^{-1}. Which of the following statements is likely to be true?

A Listeners at A and C can hear all notes below 400 Hz clearly.
B Listeners at B can only hear frequencies above 400 Hz.
C Listeners at A and C can hear all frequencies clearly.
D Listeners at A and C can only hear frequencies above 400 Hz.

Give clear reasons for your answer; include numerical values. [4 marks]

Loudspeakers for domestic sound systems are generally connected 'in phase' – the loudspeaker cones move in the same direction with the same input. That is, given the same input, they both produce compressions at the same time. They can be connected 'out of phase' so that they move in *opposite* directions. This means that, with the same input, one speaker will produce a compression whilst the other will produce a rarefaction.

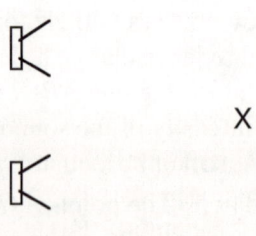

Question 158
If two speakers are connected 'out of phase', the sounds from one speaker are likely to 'cancel' the sound from the other for a listener at X. X is the same distance from each loudspeaker. Explain why. [4 marks]

4 Diffraction and interference

Two loudspeakers spaced 1.2 m apart (as shown) are both giving out a steady frequency of 400 Hz. Junia is sitting at the point A, 1.6 m from one loudspeaker.

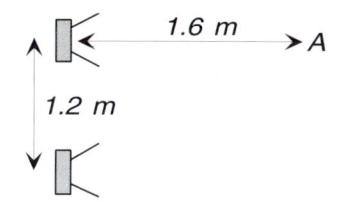

Question 159
Will the point A be a point of maximum or minimum intensity? Justify your answer. The speed of sound can be taken as 320 ms^{-1}. [4 marks]

Question 160
The frequency of the sound is changed to 800 Hz. Which of the following is correct about the point A?

A   It is a point of maximum intensity.
B   It is a point of minimum intensity.
C   It is a point of medium intensity.
D   It is a point of refractive interference.

Question 161
It is possible to improve the angular 'spread' of sound from a loudspeaker by covering the speaker with a thick piece of board with a *small* hole in its centre. The method works better for lower frequencies than for very high frequencies. Explain why this method improves the angular 'spread' of sound.

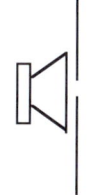

Question 162
Estimate the hole size to ensure good spread of sound at a frequency of 6500 Hz. Take the speed of sound to be 325 m s^{-1}. Show your working. [4 marks]

Question 163
A listener to a new loudspeaker notices that the spread of sound from the high frequencies is considerably less than the spread of sound from low frequencies. Explain why this is so. Your answer should include estimates of audible low-frequency wavelengths and estimates of audible high-frequency wavelengths.
[4 marks]

Question 164
Sound barriers at the side of freeways are more effective at reducing the noise of high-frequency sounds (for example 2000 Hz police sirens) than low-frequency noises (for example 200 Hz noises from trucks). Explain why this is so.

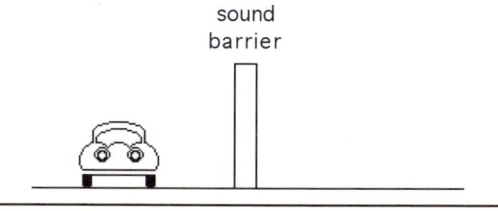

[3 marks]

4 Diffraction and interference

Two small loudspeakers are set up at one end of a school oval as shown below. Both are connected to the same source and emit sound with the same amplitude with a frequency of 1.7 kHz. The point Z is an equal distance from both loudspeakers. The speed of sound is 340 ms^{-1}.

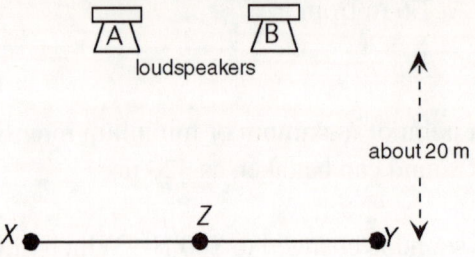

A student who walks from X to Y notices that the sound intensity decreases then increases every few steps.

Question 165 (1996 CAT)
Explain why the sound intensity varies along the line XY. You may use diagrams to illustrate your explanation. [4 marks]

Moving a distance of 1.5 m from Z towards Y, the sound intensity goes from maximum at Z through a minimum to another maximum at Y.

Question 166 (1996 CAT)
If the student is now exactly 20.4 m from speaker A, what is the distance to speaker B? Show your working. [3 marks]

In an experiment conducted in a classroom, two speakers A and B are placed apart and connected to the same audio oscillator. A sound level meter L is moved along a line which is 1.60 metres from the line between the speakers. L is initially placed directly opposite B. The frequency of the oscillator is increased from a low value, and it is found that a *minimum* intensity is recorded by L when the frequency reaches 425 Hz. The speed of sound in the classroom is 340 m s^{-1}.

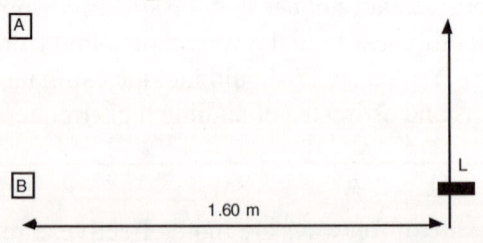

Question 167
What wavelength sound is produced by the speakers at this frequency?

Question 168
Explain carefully why the minimum occurs at this frequency.

Question 169
As the frequency of sound is increased beyond 425 Hz, calculate the frequency value for which the next intensity minimum will occur at L.

4 Diffraction and interference

Question 170
At this new value, the meter is moved up along the line shown, until it is horizontally opposite speaker A as shown in the diagram. Describe what will happen to the sound intensity recorded by L due to this move.

Mary tests the response of a loudspeaker. She uses a microphone to measure the sound intensity at different positions on a circle around the speaker, as shown. She makes measurements at 10 kHz and 100 Hz.

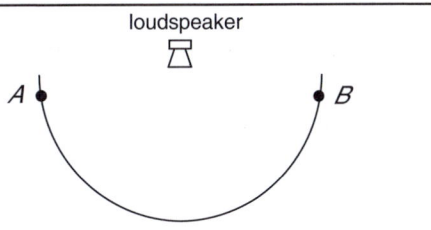

Question 171   *(1997 CAT)*
If the speed of sound in air is 340 m s^{-1}, what are the wavelengths of the waves at the two frequencies of the measurements?

The graph below shows the intensity measured for each frequency at the positions on the line shown between points A and B.

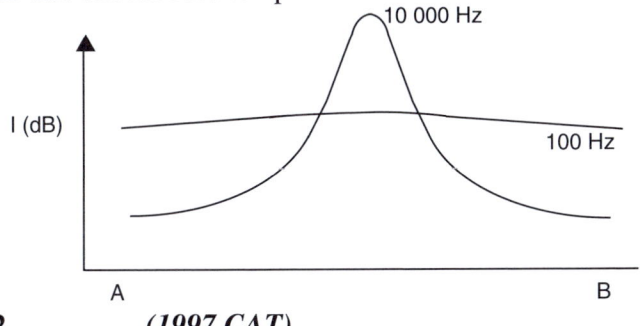

Question 172   *(1997 CAT)*
Explain why the response at 10 000 Hz is stronger directly in front of the speaker, while the response at 100 Hz is nearly the same at all positions.

[4 marks]

In order to investigate interference of sound waves, students set up the experiment illustrated below. Two small loudspeakers are connected in phase to an audio-frequency oscillator. The speakers (S1 and S2) are separated by a distance that is greater than 1 m and less than 2 m. A microphone, M, is placed 2.00 m directly in front of speaker S1. The output of the microphone is connected to an oscilloscope so that the sound intensity can be observed.

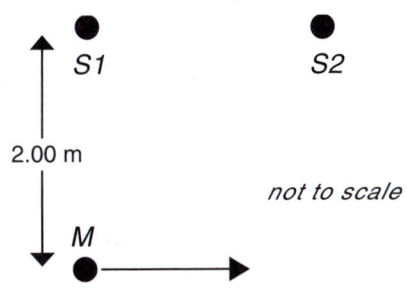

4 Diffraction and interference

The frequency of the oscillator is adjusted, and a series of maximum and minimum intensities are detected at the microphone. The lowest frequency at which an intensity minimum is detected at the microphone is 340 Hz, which corresponds to a wavelength of 1.00 m.

Question 173 *(1997 CAT)*

Explain why this minimum occurs and calculate the exact spacing between the two speakers. [4 marks]

With the frequency of the oscillator still set to 340 Hz, the microphone is then slowly moved a distance of 0.75 m to the right, along the line shown.

Question 174 *(1997 CAT)*

How many more intensity minima would be detected?

Rose is performing a sound check on speaker placement for an open air concert. Two speakers S_1 and S_2 are placed on a stage as represented in the diagram following, and she is checking the loudness for sound of a single frequency coming from the speakers.

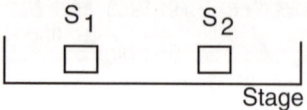

Question 175

Sitting in chair X she finds the sound is quite loud. X is at a maximum of the interference pattern caused by the two speakers. Explain why the sound intensity is a maximum at X at all times. [4 marks]

As she moves along the front row of seats, the sound intensity drops to a minimum at Y. Remembering her physics, she investigates further by measuring these distances: $S_1X = 8.7$ m; $S_2X = 10.3$ m; $S_1Y = 8.0$ m, $S_2Y = 9.2$ m.

Question 176

What is the path difference for sound from S_1 and S_2 reaching each of the points X and Y?

Question 177

What is the wavelength of sound waves from the two speakers? Show your working.

Question 178

In a further investigation Rose doubles the frequency of sound from the speakers. She finds that at chair Y the sound is now a maximum in the interference pattern. Explain why this is so. [4 marks]

4 Diffraction and interference

Dan is listening to a ship whistle coming from the docks, trying to get his bearings.

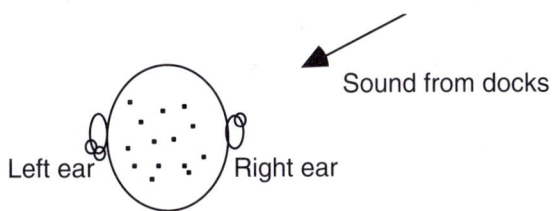

Experiments which model the head as a sphere of diameter 0.2 m, show that the sound 'shadow' caused by Dan's head reduces the intensity of the sound at his left ear compared to his right ear. This is one way Dan can tell which direction the sound is coming from. They also show that if the sound was of a lower frequency (e.g. from a foghorn) then the difference in sound intensity at Dan's left and right ears would be much less, making it harder for him to tell which direction the sound is coming from.

Question 179
Explain why, for a foghorn sound (wavelength ~ 5 m), the sound intensity at Dan's left ear would not be much less than for his right ear, whereas for a higher pitched sound (wavelength ~ 10 cm) there would be a considerable intensity difference.

[4 marks]

The other way Dan can tell which direction sounds are coming from is that sound from the whistle has to travel further to reach his left ear, and arrives later. The graph below shows the pressure variation at Dan's right ear for a sound of a single frequency.

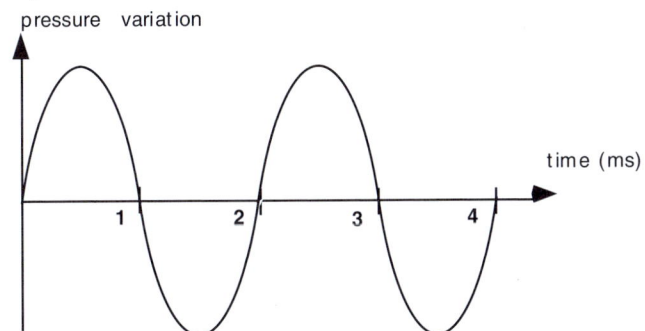

Question 180
What is the frequency of this sound?

Question 181
Superimpose on this graph a sketch of the pressure variation at Dan's left ear, assuming the sound has to travel a further 17 cm compared to the right ear. Take the speed of sound as 340 m s^{-1}. Show your reasoning.

[4 marks]

4 Diffraction and interference

Two loudspeakers (S1 and S2) placed on a football field, as shown to the right emit sound of the same frequency, in phase. Point P is at a minimum of the interference pattern produced by the sound waves.

Question 182   (1998 CAT)
Explain why the sound intensity at P is a minimum at all times.

A student walks from P towards Q. The student observes that the sound increases in intensity until it reaches a maximum at point Q.

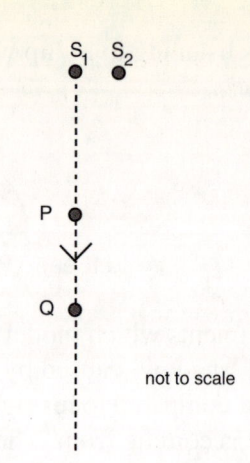

Point P is at a distance of 78.8 m from S_1 and 81.3 m from S_2. Point Q is at a distance of 99.0 m from S_1 and 101.0 m from S_2.

Question 183
What is the path difference for sound reaching point P from S_1 and S_2?

Question 184
What is the wavelength of the sound waves emitted by the two speakers? Show your working.

A teacher sets up a loudspeaker system on the school oval with two speakers placed at points X and Y as shown in the diagram following.

The speakers emit sound in phase with a wavelength of 0.20 m. A student walks from point P to point R and notices that the sound intensity is a maximum at P, then decreases to a minimum at Q and increases to another maximum at R. The distances XP and YP are equal.

Question 185   (1999 CAT)
If the distance from Y to Q is 9.0 m, what is the distance from X to Q? Show your working clearly.

[4 marks]

4 Diffraction and interference

John and Anna sat in the front row at a trumpet concert. John's seat was directly in front of the trumpet. Anna's seat was to one side, as shown. After the concert John commented on how good the sound was.

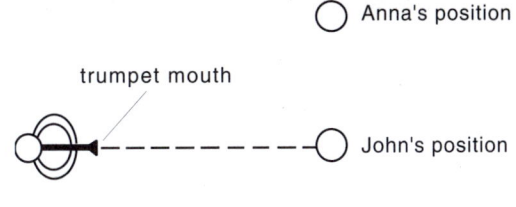

Anna noted that the high-frequency notes were relatively much weaker than the low-frequency notes.

Question 186   *(1999 CAT)*

Which of the following wave phenomena best explains their observations?

   A   reflection   **B**   refraction
   C   resonance   **D**   diffraction

[1 mark]

Question 187   *(1999 CAT)*

Explain why the sound quality was poorer where Anna was sitting than where John was sitting.

[3 marks]

Models of human heads used by sound engineers contain a microphone at the position of each ear, simulating stereo hearing. The diagram shows such a model being used to record sound from a loudspeaker some distance away. Sound reaching the left ear of the model travels 0.20 m further than sound reaching the right ear. The speaker emits sound with $\lambda = 0.80$ m.

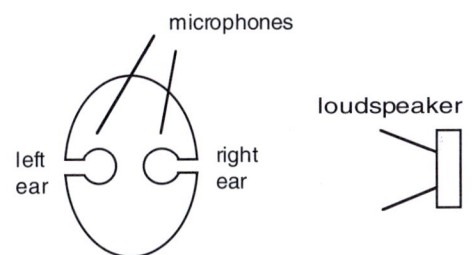

The graph shows pressure variation at the right ear as a function of time.

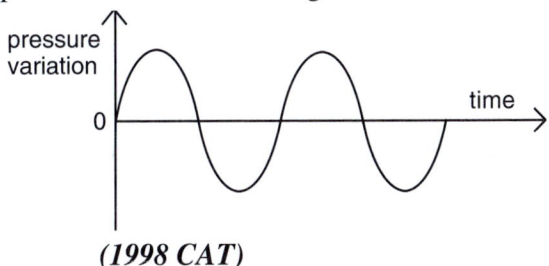

Question 188   *(1998 CAT)*

In the graphs following, the dashed curve is the variation of pressure at the right ear of the model head, and is shown for reference. Which of the graphs following best shows the pressure variation as a function of time at the left ear of the model head?

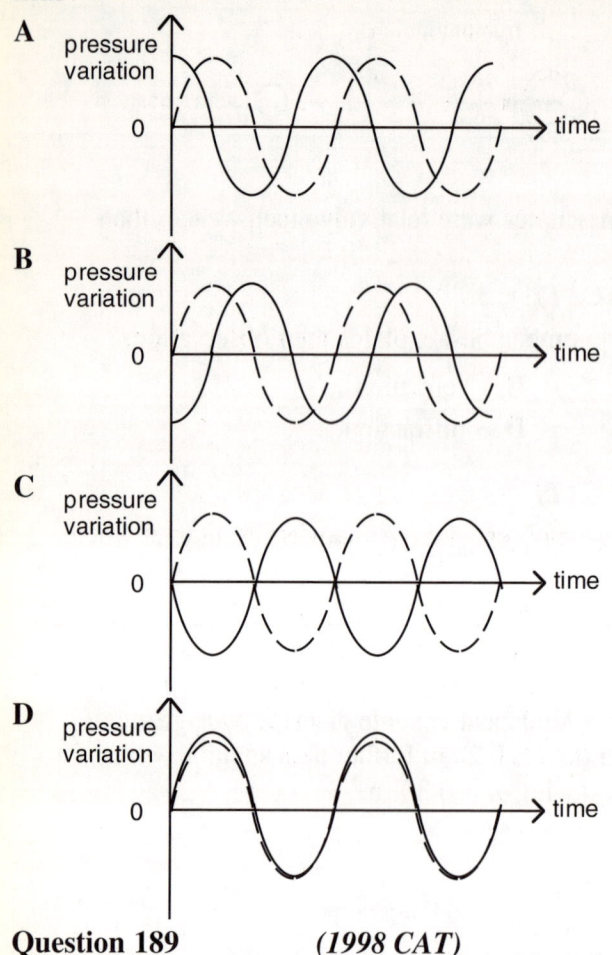

A
pressure variation

B
pressure variation

C
pressure variation

D
pressure variation

Question 189   *(1998 CAT)*
The frequency of the sound source is increased significantly to 4000 Hz. The recording engineer then notices that the sound intensity at the microphone in the left ear has dropped noticeably.
Explain why this has occurred. [3 marks]

Dan is in a nightclub listening to an acoustic band with double bass, violin and guitar. He is unfortunately seated behind a solid stand, but can arrange his seat so he can see the double bass or violin, but not both. He gets the best sound if he chooses to see the violin, since the double bass can still be heard clearly in that case, whereas the violin sounds muffled if he cannot also see the violinist.

Question 190
Explain why Dan can hear the double bass clearly when he cannot see it because of the stand, but the violin sounds muffled unless he can see it.
[4 marks]

4 Diffraction and interference

A teacher sets up two loudspeakers on the school oval. They are both connected to the same signal generator, with cables of the same length, and are connected in phase.

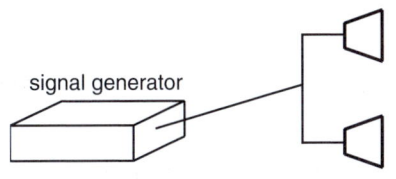

Xena hears a pure tone from the loudspeakers, and notices that the sound intensity is different depending on her location on the oval. The teacher tells Xena that the frequency of the tone is between 150 and 250 Hz, but does not tell her the exact value.

Xena notices that the sound is loudest at points at equal distances from the two loudspeakers. In moving around, there are several places where the sound level is a *minimum*. One of these places is 2.4 m further from one loudspeaker than from the other.

Question 191 *(2001 written examination)*

What is the frequency of the sound from the loudspeakers? Show your working and explain your reasoning. (Take the speed of sound to be 340 ms^{-1}.)

[4 marks]

John and Maria are discussing how we determine the direction from which a sound comes. The setup they are considering is shown below. John says that we can tell the direction of the sound from the speaker because the brain can detect the time difference between the arrival time of the peak of a wave at each ear. Maria says that this cannot be so: we tell the direction of sound because diffraction causes the intensity at the more distant ear to be lower, and the brain detects this difference.

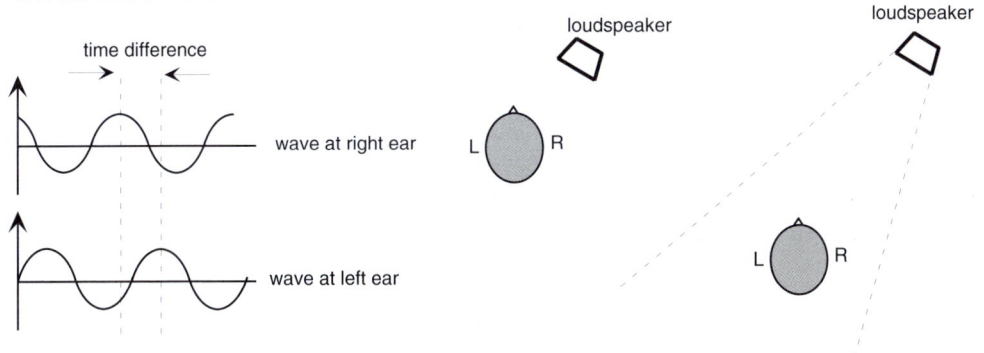

Question 192 *(2001 written examination)*

Discuss to what extent, and under what conditions, their explanations are correct. You may use diagrams to explain your answer.

[4 marks]

CHAPTER 5 — Magnetic and electric basics

An electric motor is being supplied with a current of 15 A. Electric power is being continuously supplied to the motor at the rate of 225 W.

Question 193

How much energy is used by the motor in 1 hour of continuous operation?

Question 194

What is the value of the ratio (the effective resistance)

$$\frac{\text{voltage across the motor terminals}}{\text{current through the motor}} \ ?$$

Question 195

How much charge flows into the motor in one hour?

A motorist turns on a car's headlights and then, with the lights still on, 5 seconds later turns on the starter motor. The headlights and the starter motor are connected in parallel. The battery voltage is always 12 V. The diagram below shows the electric power supplied by the battery as a function of time.

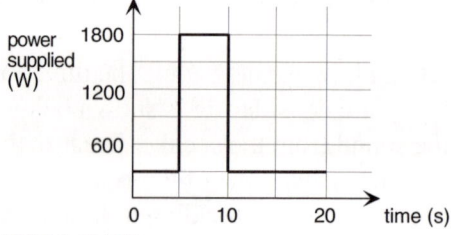

Question 196   *(1996 CAT)*

Assuming all the electric energy is supplied by the battery, how much energy does the battery supply in the first 15 seconds?

Question 197   *(1996 CAT)*

What is the current in the starter motor while the motor is turned on?

Question 198   *(1996 CAT)*

How much electrical energy does the battery supply to each coulomb of charge flowing through the starter motor?

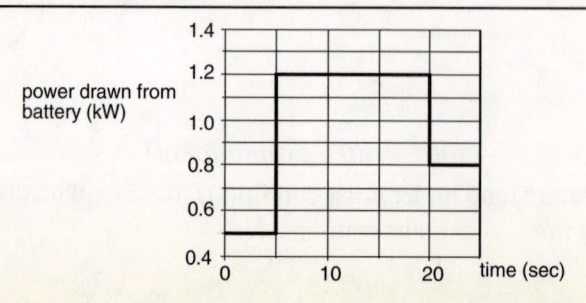

5 Magnetic and electric basics

Question 199
The power supplied by a 24 V car battery is graphed above. How much energy is drawn from this battery between t = 5 & t = 15 s?

Question 200
How much current is drawn at time t = 7 seconds?

Question 201
How many coulombs pass through the battery between t = 5 s & t = 20 s?

A 120 V farm generator is supplying current to a number of appliances. The circuit below models the arrangement of the appliances.

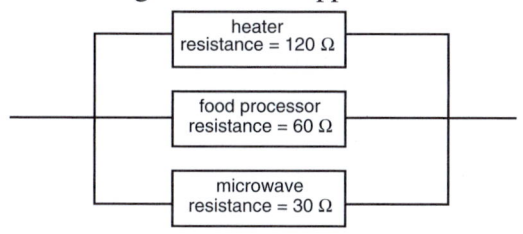

Question 202
What is the total resistance of the three appliances connected as above?

Question 203
What power is drawn by this arrangement when both the generator and the appliances are operating normally?

Question 204
What charge flows if both the generator and appliances operate normally for 24 hours?

Question 205
What energy is used if both the generator and the appliances operate normally for 24 hours?

Question 206
If the three appliances were connected in *series*, what current would flow through each of the them? *(Assume they have the same resistance.)*

Question 207
In this arrangement, would any of them operate normally? Give reasons.

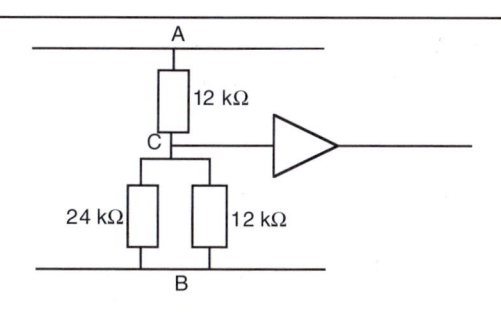

[4 marks]

5 Magnetic and electric basics

Question 208
The input to a NOT gate is connected as shown in the circuit above. What is the total resistance of the three resistors between A and B?

Question 209
If the 24 kΩ resistor were removed from the arrangement, would the voltage at C rise or fall? Justify your answer with numerical calculations. [4 marks]

Anna is investigating the force on a current carrying wire, due to a solenoid coil wrapped around a metal bar. The arrangement is shown below. Current flows through the wire wrapped around the metal bar.

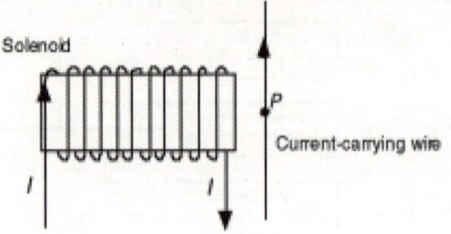

Question 210
Which of the following best describes the direction of the magnetic force on the current carrying wire at the point *P*, as shown in the diagram above?

A   into the page   B   out of the page   C   vertically up
D   to the right   E   to the left   F   zero

Electrons moving through a resistor experience a force if the resistor is in a magnetic field. A resistor is oriented east–west as shown, with the earth's magnetic field into the page. The current is to the right.

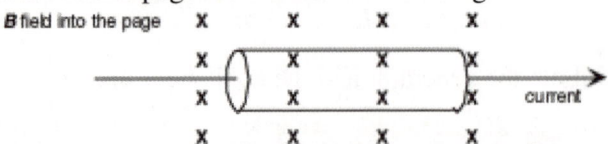

Question 211
An electron travelling through the resistor will experience a force due to the magnetic field which is:

A up   B down   C into the page   D out of the page   E zero

Question 212
A (vertical) lightning conductor gets a 'bolt from the blue'. This sends *electrons* in the conductor rushing downwards. Assume the earth's magnetic field is horizontal.

In which direction does the earth's magnetic field push the lightning conductor?

A   North   B   South   C   East   D   West

5 Magnetic and electric basics

Question 213
If the current in the conductor is of the order of 10^4 A, estimate the size of the force on the lightning conductor.

During a severe storm, the lightning conductor is tipped over towards the north, so that it now makes an angle of 30° to the horizontal, and is no longer at right angles to the magnetic field.

Question 214
Which of the following best describes the new magnetic force on the lightning conductor when a current of around 10^4 A flows through it?

A   It is the same as before.
B   It is greater than before.
C   It is less than before, but not zero.
D   There is now no magnetic force on the conductor.

Question 215
A wire of length 2 m in a magnetic field of 0.15 T experiences a sideways magnetic force of 3 N. What is the current flowing through it?

The coils providing the magnetic field for an electric motor are sketched to the right. A strong current is flowing through both coils. As a result they exert forces on each other.

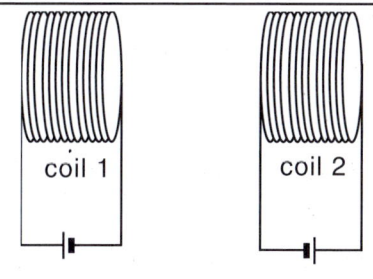

Question 216
What is the direction of the force on the left hand coil (coil 1)? Give reasons.

[4 marks]

Question 217
When a wire is pushed by a magnetic force perpendicular to the magnetic field, which *one or more* of the following can you be sure about?

A   A current is flowing in the wire.
B   The force F on the wire is given by the formula $F = BI$.
C   The force on the wire is equal to the sum of all the magnetic forces on the moving electrons inside the wire.
D   The wire is in the same direction as the magnetic field.

Jon is investigating fields produced by magnets as background research for a model electric motor. One arrangement of magnets is shown. The magnets are identical; *P* is exactly midway between the two north poles.

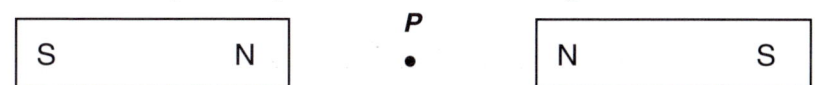

5 Magnetic and electric basics

Question 218
Which of the following best describes the magnetic field strength at *P*?

A   It is stronger than the field due to one magnet only.
B   It is about the same as the field due to one magnet only.
C   It is weaker than the field due to one magnet only.
D   It is close to zero.

An AC motor consumes 16 kW (RMS) of power. Its resistance is 11 Ω.

Question 219
What is the RMS value of the AC current?

Question 220
What is the RMS value of the AC voltage?

Question 221
What is the peak value of the AC current?

Question 222
What is the peak value of the AC voltage?

Question 223
A CD player draws a current of 0.15 A (RMS) from 240 V (RMS). Calculate its power consumption.

Question 224
Calculate how many joules of energy the player uses in 2 hours of playing.

Question 225
An electric heater runs on 400 V (peak) and 10 A (peak). How many joules of electrical energy does it consume (on the average) every second?

Question 226
Stan buys a night lighting system overseas. It is designed to run on an RMS voltage of 80 V, with a power rating of 40 W. What is the RMS current in the lighting system when it operates at its stated rating?

He intends to buy a transformer to operate the system, but a friend suggests wiring in a resistor *R* to drop the voltage from the 240V RMS mains value in Australia, using the circuit shown below.

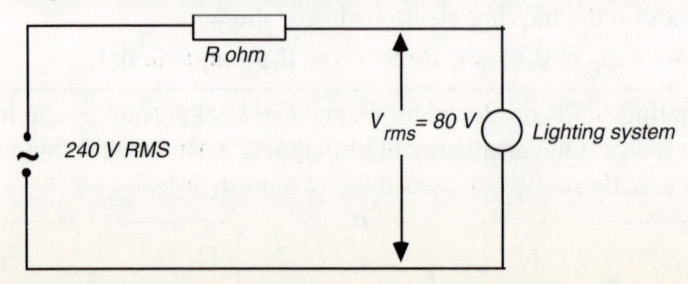

5 Magnetic and electric basics

Question 227
In the circuit above, what would the value of *R* need to be, to ensure 80 V across the lighting system?

Question 228
For the circuit above, what would be the power supplied from the mains?

Question 229
Explain why this method is inferior to using a step-down transformer.
[4 marks]

A generator supplies electricity to two distant motors in a workshop. The voltage at the generator is 230 V (RMS). When both motors are operating, the voltage at the workshop is 220 V (RMS); the current in motor 1 is 13 A (RMS); in motor 2 it is 10 A (RMS). The sketch below shows the generator connected to the workshop (the wiring to the motors has been omitted).

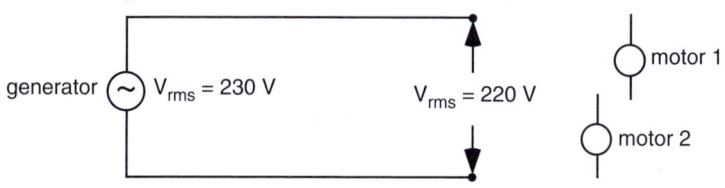

Question 230   *(1997 CAT)*
Show the wiring to the two motors.

Question 231   *(1997 CAT)*
How much power is supplied by the generator when the two motors are running?

Question 232   *(1997 CAT)*
How much electric power is lost in the cables which connect the generator and the workshop? Show your working.   [4 marks]

Question 233
A solenoid consisting of 10 coil windings is shown following. Which of the four magnetic field patterns A–D would be formed in the shaded plane shown, viewed from above the plane? Note that **x** means current flowing *downwards*, • means current flowing upwards.

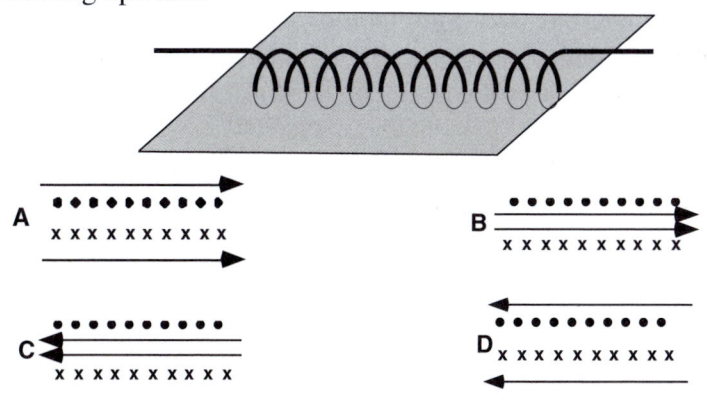

5 Magnetic and electric basics

Question 234 *(1998 CAT)*

Diagrams A, B and C show three sources of magnetic fields. Diagrams D, E and F show three possible magnetic field patterns across the plane indicated by the dashed square. For each source identify the correct field pattern.

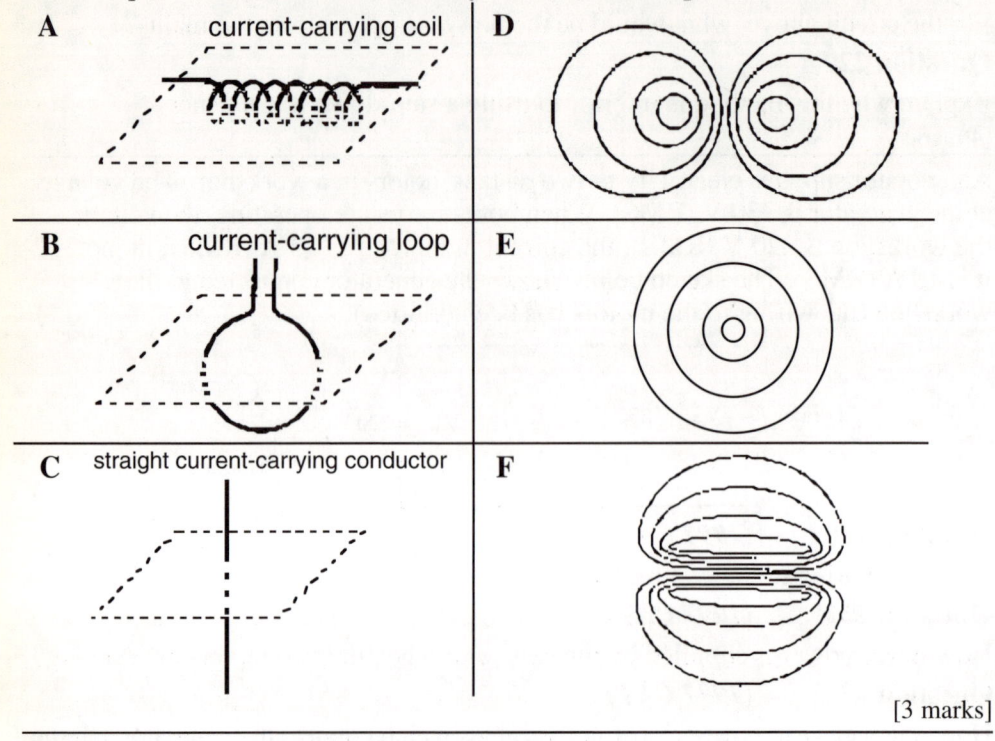

A	current-carrying coil	D	
B	current-carrying loop	E	
C	straight current-carrying conductor	F	

[3 marks]

Question 235 *(2000 written examination)*

A lightning discharge between the ground and a cloud is composed of a series of strokes. The duration of each lightning stroke is typically 30 µs. The power per stroke is 1.0×10^{12} W. If the voltage between the cloud and the ground is 4.0×10^6 V, how much charge is transferred in a single stroke?

[4 marks]

Question 236 *(2000 written examination)*

The current in the lightning stroke passes from ground to cloud. The result of this is to generate a magnetic field in the region of the stroke. The direction of the current I is shown. The field at X is shown as an arrow.

Which *one* of the diagrams below best indicates the *direction* of the magnetic field at point X, a distance r from the lightning stroke?

5 Magnetic and electric basics

A — the field is a tangent to the circle

B — the field is normal to the circle

C — the field is radial to the circle

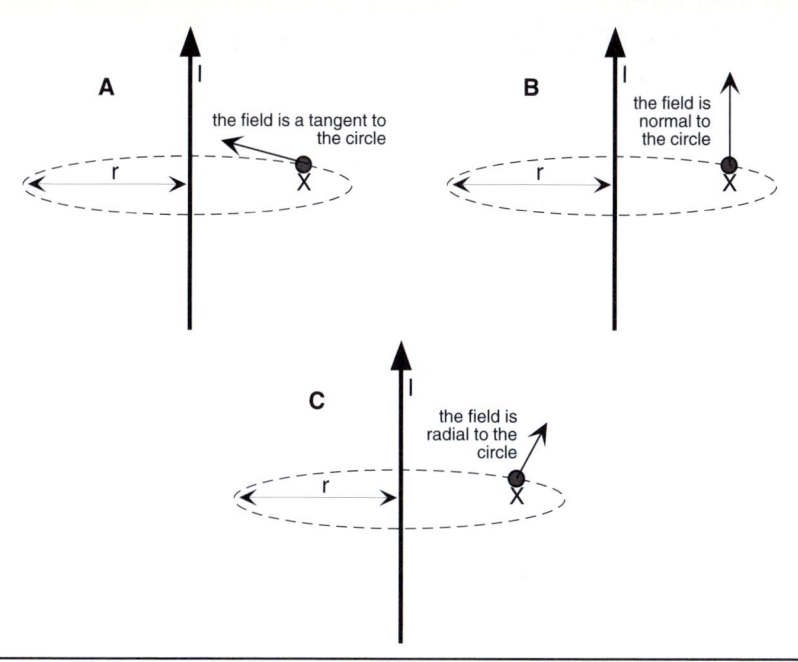

A portable electric heater has two settings, 'high' and 'low'. These heating levels are obtained by connecting two heating elements either in series or in parallel, across the 240 V_{RMS} mains supply. When the heating elements are connected in series, the total power dissipated in them is 960 W.

Question 237   *(2001 written examination)*

What is the resistance of each element?

Question 238   *(2001 written examination)*

What is the value of the ratio

$$\frac{\text{total power dissipated in heating elements in series}}{\text{total power dissipated in heating elements in parallel}}?$$

A student is provided with several lengths of wire, plus the following electrical components:

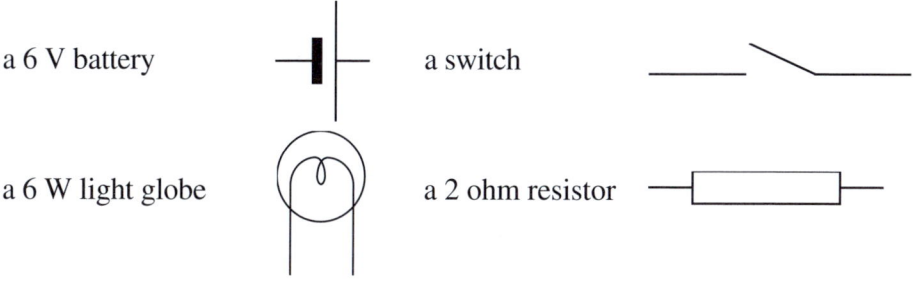

a 6 V battery

a switch

a 6 W light globe

a 2 ohm resistor

Question 239   *(2001 written examination)*

Sketch a circuit that could be connected using all the components, so that the globe lights up all the time, but becomes brighter when the switch is closed.

5 Magnetic and electric basics

A DC current of 100 mA flows in a thick copper wire. The voltage across the ends of the wire is 1.0×10^{-4} V.

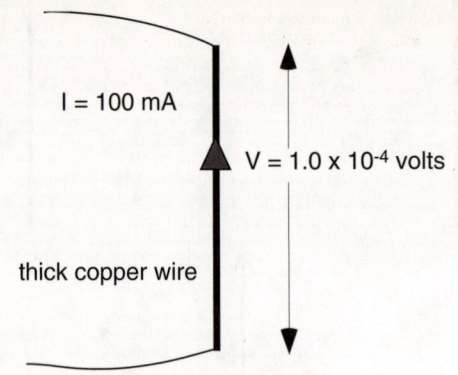

```
            I = 100 mA
                        V = 1.0 x 10⁻⁴ volts

            thick copper wire
```

Question 240   *(2001 written examination)*
What is the resistance of the thick copper wire?

Question 241   *(2001 written examination)*
How many electrons enter the copper wire each second? (An electron has a charge of 1.60×10^{-19} C.)

CHAPTER 6   Motors

A simple DC motor, like the one shown below, has one coil. Its field is supplied by permanent magnets.

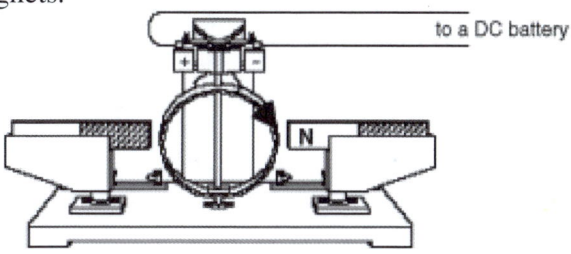

The direction of the current in the coil is shown by the arrow on the coil – clockwise in the diagram. The shaded part at the end of each magnet is a *south* pole, and the clear part of each magnet is a *north* pole.

Question 242

When current flows through the coil, the motor will, *viewed from above the motor:*

A   rotate clockwise.
B   rotate anti-clockwise.
C   rotate either way, depending on the initial direction of motion.
D   not rotate, it will simply vibrate about the position shown.

Question 243

The arrangement at the top of the diagram where the current from the battery is transferred to the rotating coil is called the *commutator*. Which one or more of the following would be likely to decrease the speed of the motor? *(You may assume that the rotating coil is very light.)*

A   Increase the number of turns of the rotating coil.
B   Decrease the number of turns of the rotating coil.
C   Increase the area of the rotating coil.
D   Decrease the area of the rotating coil.
E   Increase the current in the rotating coil.
F   Decrease the current in the rotating coil.

Question 244

If the permanent magnets were replaced by DC electromagnets with the same poles, which of the following changes would be necessary to keep the motor rotating as above?

A   Replace the commutator with slip rings.
B   Rectify the DC current.
C   Reverse the direction of the poles.
D   Make no changes.

6 Motors

The sketch shows a coil on an axle between the poles of an electromagnet.

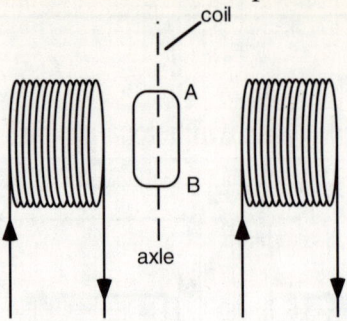

A current is flowing in the coil between the electromagnets. The coil is rotating like a motor. Side AB of the coil is rotating towards us.

Question 245
Which of the following best describes the current in the coil? Justify.

A   It is flowing from A to B along this side.
B   It must be alternating.
C   It is flowing from B to A along this side.
D   It could be flowing in either direction, one cannot tell.   [3 marks]

Jimmy Spriggs thinks that reversing the current flow in *one* of the electromagnets in the arrangement above could increase the magnetic field and hence the speed of the motor.

Question 246
Do you agree that this would increase the magnetic field? Give reasons.
[3 marks]

A small electric motor, as shown, runs off DC. Two *electromagnets*, as shown, supply the magnetic field for the stator (the field coils).

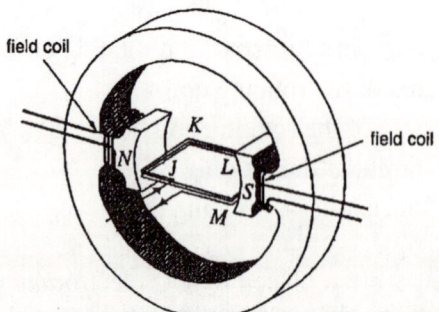

The same current source is also used to supply the rotor (the rotating coil). The field coils are connected in *series* with the rotating coil. An inquisitive student connects the motor sketched above to a source of *AC* current. The field coils and the rotating coil are still connected in series. She expects to see the motor just vibrate, but not rotate. To her surprise, it rotates just as it did for the DC current input.

6 Motors

Question 247
Explain why a DC source and an AC source cause the same effect in this motor.

[4 marks]

A book describes a simple motor made from a coil of copper wire resting on supports made out of paper clips with a magnet placed below. The paper clips make electrical contact with the battery. The design is shown in the following diagram. The coil is meant to spin between the supports. The coil part is made of insulated copper wire but the section resting on the paper clips are free of insulation and make good contacts. The instructions say the loop must be carefully balanced if it is to rotate freely. The side *A* of the magnet is a North pole.

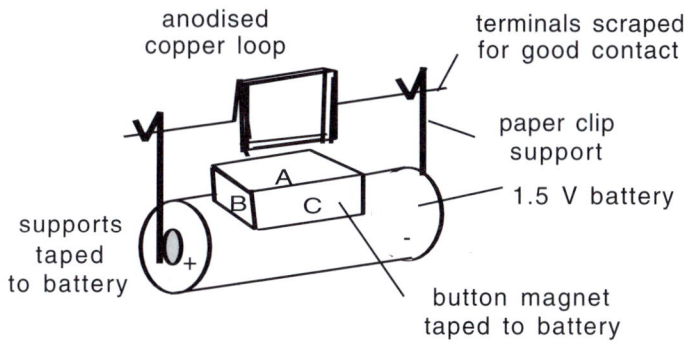

Question 248
Explain why this simple motor *should not work*. Include an explanation of what you think will happen to the coil when the connection is made.

[5 marks]

Question 249
In fact, the simple motor *does* sometimes work. Which of the following factors might explain this? Give reasons for your choice of answer.

[3 marks]

A	The battery is more effective than supposed.
B	The wiring and contact points may be better than supposed.
C	The current from the battery may switch direction.
D	The action of the motor spinning may cause 'make and break' the contact points by bouncing in a regular fashion.
E	The magnet is a specially designed one, oriented carefully.

A small electric motor comes with a toy car assembly kit. It consists of a plastic frame around which wire is wound, and thin metal sleeves that fit onto the plastic axle to act as commutator. The arrangement is shown in the diagram following. The position of the magnets is not shown.

6 Motors

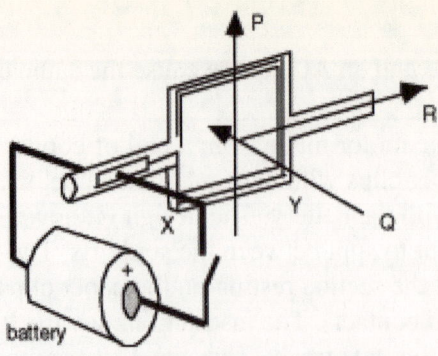

Dan is constructing the motor, following the instructions as best he can. He arranges the magnets to create a magnetic field that points up in the direction shown by *P*. With the coil vertical, as shown in the diagram, the switch is closed.

Question 250

Which of the following best describes what should happen when the switch is closed?

A   The coil should start spinning clockwise as seen from *R*.
B   The coil should start spinning anticlockwise as seen from *R*.
C   The coil will experience a net force in the direction of *Q*.
D   The coil will experience no rotational effect in this particular arrangement.

Question 251

The coil is square with 2 cm sides. It has 10 windings. The current through the coil is 4 A. The size of the field is 0.15 T. Calculate the size of the force on XY.

Question 252

If Dan directs the magnetic field along the direction *Q*, which of the above best describes the starting situation of the motor? Give reasons for your answer.

[3 marks]

The coil and permanent magnets of a DC motor are sketched in the diagram following.

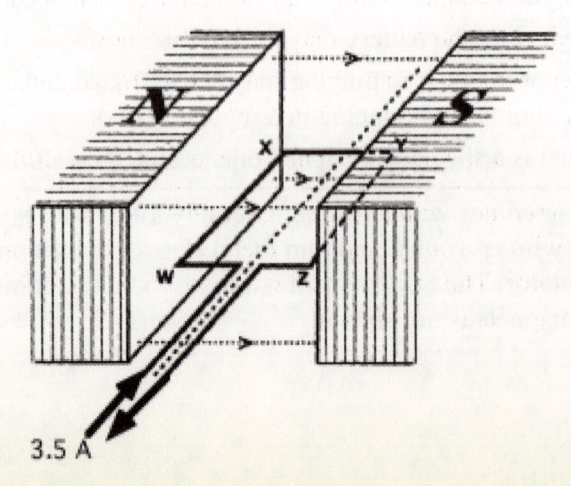

6 Motors

The magnets shown above produce a magnetic field of 45 mT between the poles shown. The coil WXYZ has 30 turns and a current of 3.5 A is flowing in each turn. The lengths WX and YZ are equal to 10 cm, and the lengths XY and ZW are equal to 5 cm.

Question 253
Sketch on the diagram the directions of the forces acting on the sides WX and ZY.

Question 254
Calculate the magnitude of the total force on the side *WX*.

Question 255
Calculate the magnitude of the total force on the side *XY*.

Simple DC electric motors can be represented in diagram form below. A flat coil is placed between the N and S poles of a permanent magnet, where it is free to rotate (the axle lies along the direction of the dashed line in the diagram below).

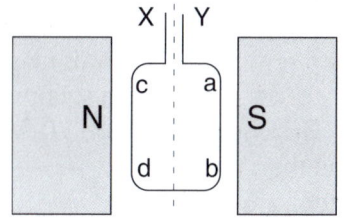

Question 256
If **X** is at a higher potential than **Y**, current will flow through the flat coil. Magnetic forces will act on the sides **ab** and **cd**. Describe the directions of these forces, and indicate whether they will result in rotation of the coil. [3 marks]

Question 257
For the coil to continue to rotate (and hence work as a motor) a *commutator* is required between the points XY and the DC supply. Explain the *function* of the commutator (you do not need to draw it). [3 marks]

A small DC motor is sketched below. It is designed to rotate vertically. It has of a coil of side 2 cm containing 20 loops of wire, situated between two electromagnets (as shown) that are shaped to give a uniform field over most of the rotation of 0.1 T. The current through the coil is 2.0 A.

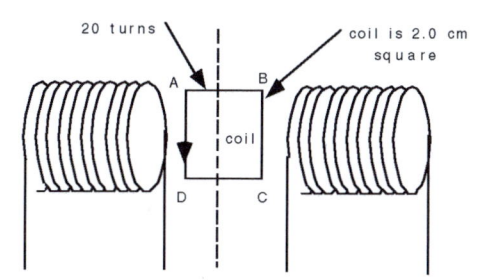

Question 258
When the coil is as shown, calculate the size of the force on side AD.

6 Motors

Question 259
When the coil is as shown, calculate the size of the force on the side AB.

Question 260
If the coil has current flowing through it as shown and it is rotating so that, *viewed from above*, it appears to be rotating clockwise, which of the following best describes the direction of the magnetic field at this time?

A	left to right	B	right to left
C	up the page	D	down the page

A DC electric motor is illustrated below.

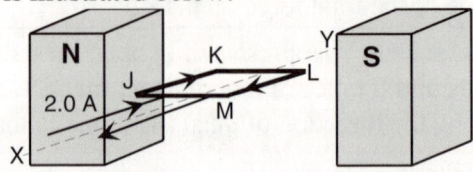

The permanent magnet shown produces a uniform magnetic field of 3.0×10^{-2} T between the poles. The coil *JKLM*, wound on a square armature of side 0.05 m, consists of 20 turns of wire. There is a current of 2.0 A in each turn of the coil. The armature can rotate about the axis *XY*.

Question 261   (1996 CAT)
For the coil oriented horizontally, as shown above, calculate the magnitude of the total force exerted on the 20 turns of side *JK*.

Question 262   (1996 CAT)
For the coil oriented horizontally, as shown in Figure 3, calculate the magnitude of the total force exerted on the 20 turns of side *KL*.

Curving the pole pieces of a magnet produces the magnetic field shown below. The maximum value of this field is 3.0×10^{-2} T. This magnetic field will provide a greater average torque on coil *JKLM*, with the same coil current, than the magnetic field shown earlier.

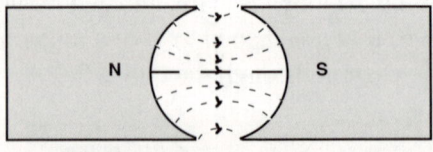

Question 263   (1996 CAT)
Explain why such pole pieces lead to a greater average torque on the coil.

(4 marks)

The diagram following shows four positions (**A, B, C, D**) of a DC motor coil. The single turn is placed in a uniform magnetic field parallel to the coil when as shown in position **A**. It rotates in the direction indicated, about the axis shown as a dashed line. The coil is attached to a commutator, to which current is passed by brushes (not shown).

6 Motors

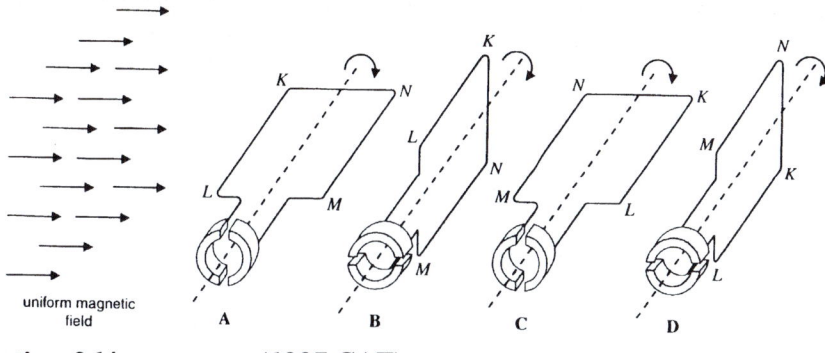

Question 264 *(1997 CAT)*

For the coil shown in position **A**, in which direction is the current flowing in side *KL*? Explain your answer.

Question 265 *(1997 CAT)*

Side *KL* of the coil is 0.10 m long. A magnetic force of 0.60 N acts on it. If the magnetic field has a magnitude of 1.5 T, what is the magnitude of the current in the coil?

Question 266 *(1997 CAT)*

Consider the two cases of the coil at rest in position **A** and the coil at rest in **B**. Explain what would happen in each case, when current flows in the coil. Discuss the forces on each side of the coil and their net effect.

[4 marks]

The figure below shows a single turn loop of wire in a uniform magnetic field. The loop's position can be altered; three orientations are shown. In each case there is a current flowing in the coil from K to L to M to N.

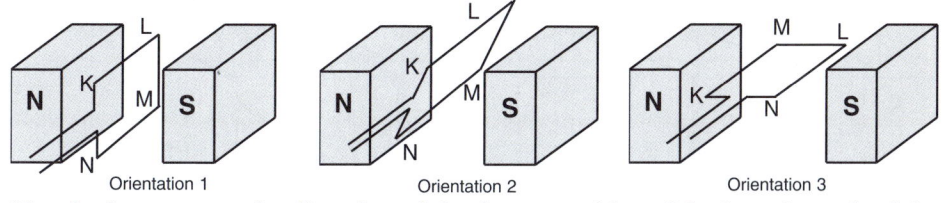

Use the key to state the direction of the force on sides of the loop in each of the orientations shown. **A**: Vertically up; **B**: Vertically down; **C**: To the left; **D**: To the right; **E**: Toward the centre of the loop; **NF**: No force

Question 267

The force on the side KL in each of the three orientations.

Question 268

The force on the side LM in each of the three orientations.

The diagrams following show a loop of wire in a uniform magnetic field. The loop can rotate and is shown at three different positions. In each case there is a current flowing around the coil from W to X to V to Z.

65

6 Motors

Question 269 *(1998 CAT)*

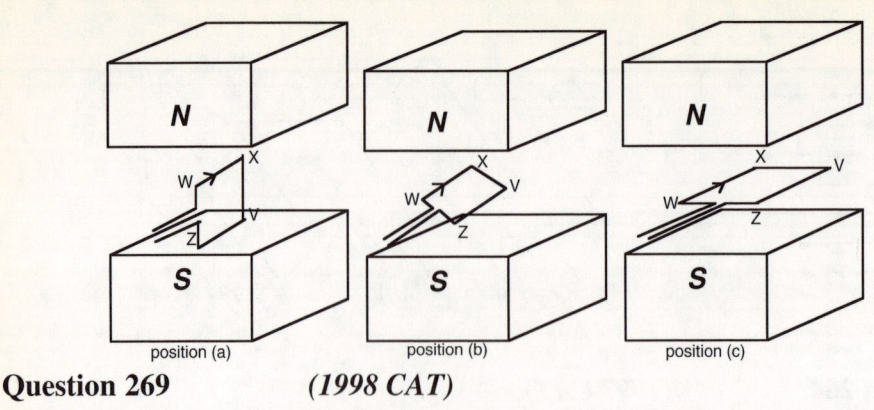

The magnetic field is 0.010 T and the current in the loop is 0.30 A. With the loop as in (a), what is the magnitude of the force acting on side WX of the coil? The length of the side WX is 0.30 m. Show your working.

In the diagram below, the arrows indicate possible directions of the force on the side WX of the loop in the three positions. The arrows in each position are in a plane perpendicular to the axis of rotation of the loop.

Question 270 *(1998 CAT)*

Which arrow in each position best represents the direction of the magnetic force on the side WX? If there is no force, write NF.

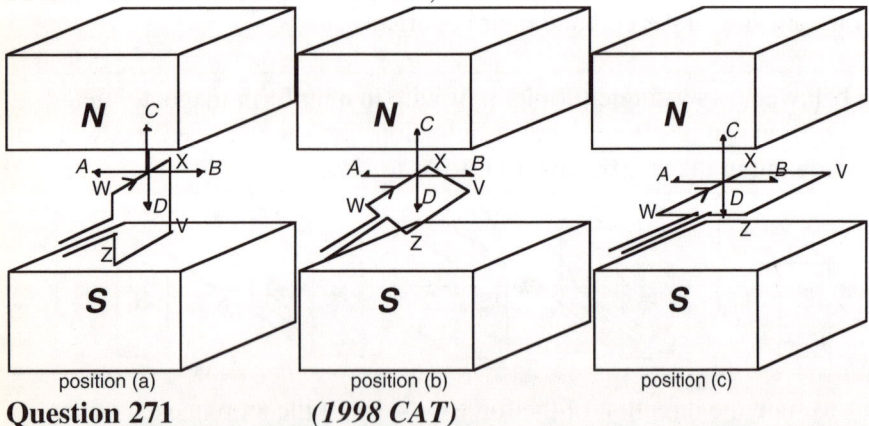

Question 271 *(1998 CAT)*

In the diagram following, the arrows indicate possible directions of the force on side XY for the loop in positions (a) and (c). For each of the two positions shown, indicate which arrow best represents the direction of the magnetic force on side XY of the loop. If there is no force, write NF.

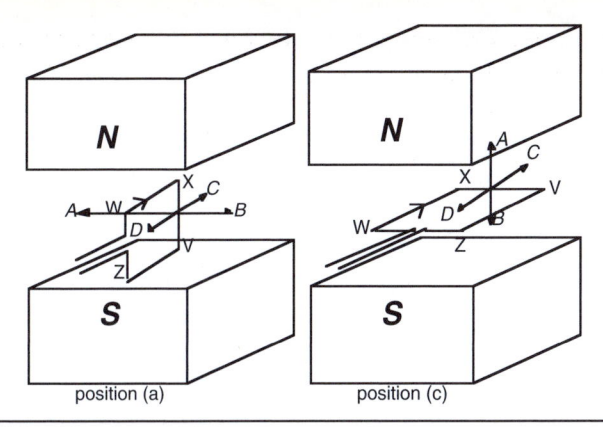

A 50-turn coil (JKLM), is free to rotate about the axis shown as the dashed line below. The coil is placed between the poles of a magnet in a uniform magnetic field of 0.040 T. The current in the coil is 1.5 A.

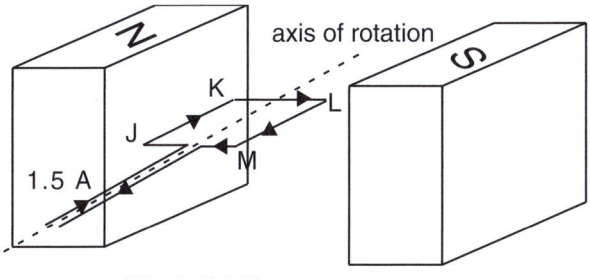

Question 272   *(1999 CAT)*

What is the size of the magnetic force on the side JK (length = 0.050 m) of the 50 turn coil when oriented as shown in the diagram above?

Question 273   *(1999 CAT)*

What is the size of the magnetic force on the side KL (length = 0.040 m) of the 50 turn coil when oriented as shown in the diagram above?

The current is now turned off, the coil is at rest, oriented as shown in the diagram below. The current is then turned on again.

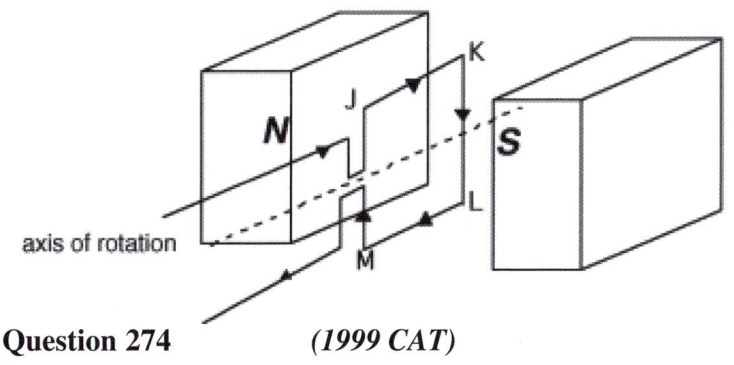

Question 274   *(1999 CAT)*

What happens to the coil after the current is turned on? Explain your answer.

[4 marks]

6 Motors

Two students examine a DC motor. They find that it has an armature consisting of rectangular coil with 50 turns, as shown below.

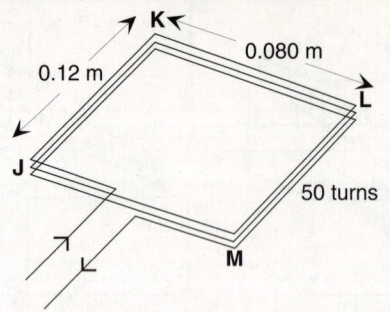

They observe that the armature is in the field of a two-pole magnet, and can rotate about an axis as shown in the diagram below. The magnetic field is produced by the current flowing through field coils. This is also shown below. The armature windings and the field coils are connected in *series*, so that the same current flows through each. The current to the armature flows through a commutator which is not shown. When the motor is operating, the current flowing is 1.5 A.

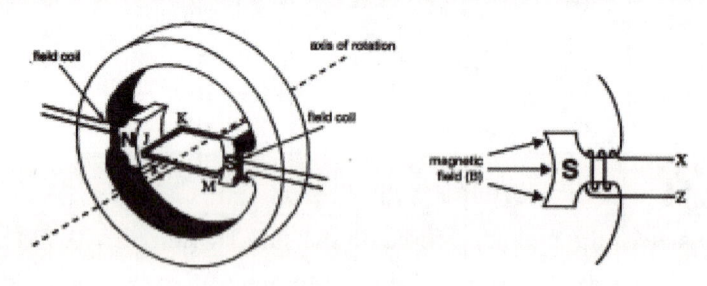

Question 275   *(2001 written examination)*

In which direction must the current flow through the field coil to produce a field as indicated by the arrows in the diagram above?

A   in at X and out at Z
B   in at Z and out at X
C   it is an AC current, so the direction is always changing
D   in either direction; the field direction does not depend on the current direction

Question 276   *(2001 written examination)*

With the armature oriented as shown in the diagram above, the magnetic field in the region of side JK is 0.10 T. A current of 1.5 A flows through the armature. What is the magnitude of the force on the side JK of the armature?

6 Motors

Question 277   *(2001 written examination)*

The armature is at rest in the orientation as shown above, when the current begins to flow in the direction shown by the arrows. Which one of the following statements best describes the initial motion of the armature?

A. It will start to rotate anticlockwise.
B. It will start to rotate clockwise.
C. It will start to rotate but the direction cannot be predicted.
D. It will oscillate about the position shown in the diagram above.

Thomas is trying to build a simple DC motor. He decides first to study the magnetic forces on a current-carrying wire. He places a single loop of wire in a uniform magnetic field, and connects a battery, as shown in the following diagram.

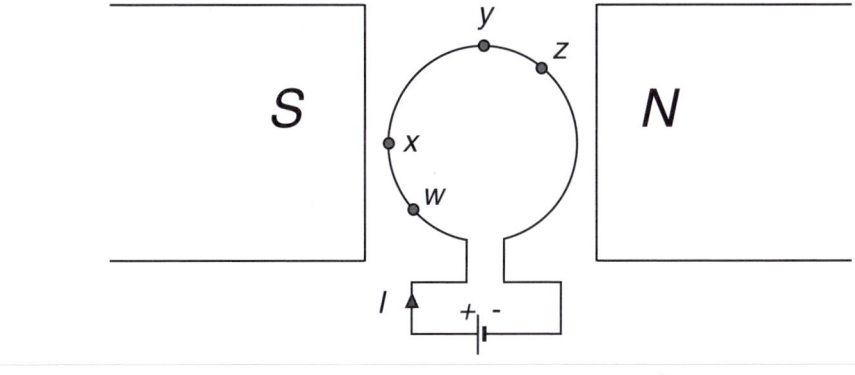

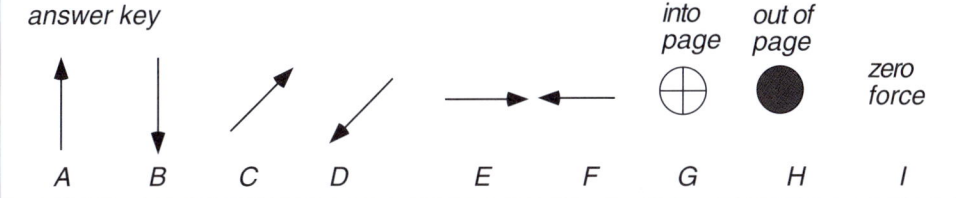

Question 278   *(2000 written examination)*

Which of the answers in the answer key best indicates the direction of the magnetic force on the wire at the point **x**?

[1 mark]

Question 279   *(2000 written examination)*

Which choice best indicates the direction of the magnetic force on the wire at the point **y**?

[1 mark]

Question 280   *(2000 written examination)*

Which choice best indicates the direction of the magnetic force on the wire at the point **z**?

[1 mark]

6 Motors

Question 281 *(2000 written examination)*

Which choice best indicates the direction of the magnetic force on the wire at the point **w**?

[1 mark]

In order to convert the above arrangement into a motor, Thomas provides an axis for rotation. He realises that there must be current flowing through the coil when it is rotating, so he attaches a set of slip rings that rotate with the coil as shown in the following diagram. The coil is initially set with its plane parallel to the magnetic field as shown, and the switch is open so that no current flows.

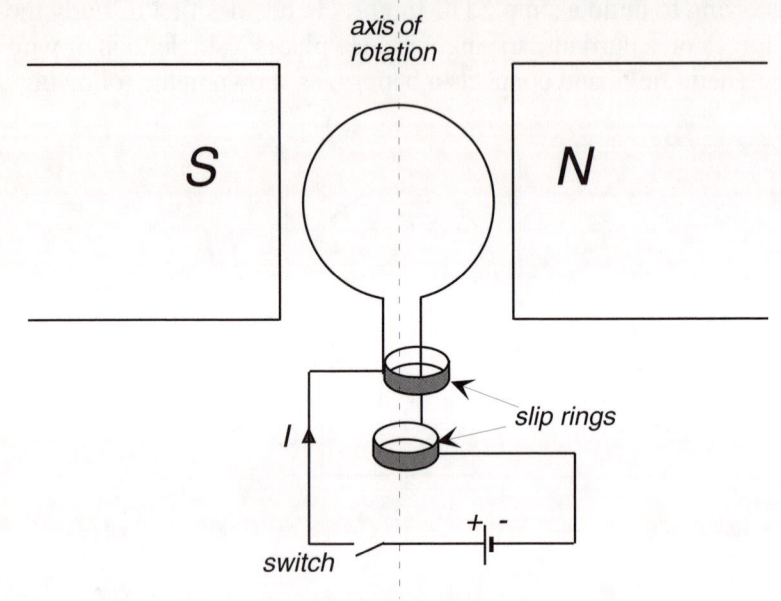

Question 282 *(2000 written examination)*

The switch is closed. Which one of the statements below best describes the situation after the current begins to flow?

A   The coil begins to rotate, but stops after turning through 90^0.
B   The coil begins to rotate and, after rotating one half turn, rotates back to its original position. It then continues to oscillate in this way.
C   The coil does not move.
D   The rotates continuously.

Question 283 *(2000 written examination)*

If the slip rings used by Thomas in the circuit above were replaced with a commutator, which one of the options in the previous question best describes the situation after the switch is closed, and the current begins to flow?

CHAPTER 7 — Generation principles

Model generators can be constructed by rotating a simple coil with lots of turns between the poles of a horseshoe magnet. This magnet produces an approximately constant magnetic field. The coil needs to be connected to an external circuit to make use of the electricity. This can be done by 'slip rings'. The coil is turned at a slow steady speed and the voltage between the points A and B is observed with a C.R.O. An AC signal with a peak value of 1.5 V and a period of 2.5 s is observed.

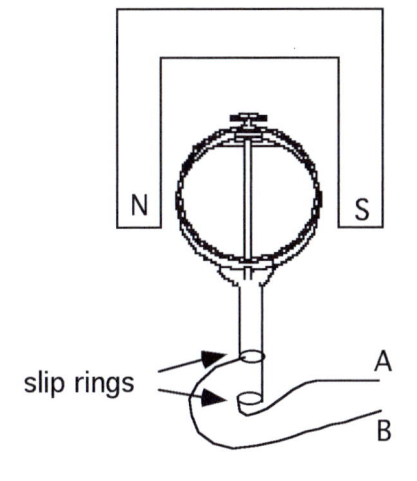

Question 284
Sketch this voltage. Label the axes with correct numerical values.

[3 marks]

Question 285
Explain why an *AC* voltage is produced. Explain why the voltage is *not* DC. You should refer to the basic principle of physics involved.

[5 marks]

Question 286
As the coil is rotated through a complete turn, which of the following statements about the *flux through the coil* is correct?

A   It remains constant.
B   It varies from zero to a positive value and back to zero.
C   It varies from zero to a positive value, to zero, to the positive value and then back to zero.
D   It varies from zero to a positive value, to zero, to a negative value and then back to zero.

Question 287
A magnet drops through a coil of area 0.2 m^2. Which of the graphs below best describes how the flux through the coil changes with time?

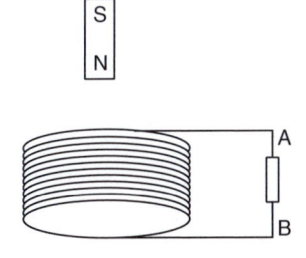

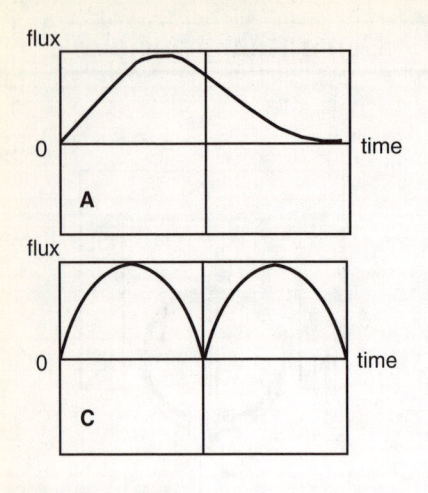

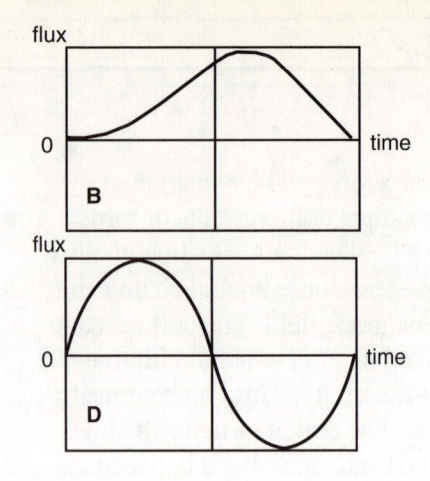

7 Generation principles

A flux vs time: rises then falls (single hump, asymmetric)

B flux vs time: rises then falls (single hump, symmetric)

C flux vs time: two positive humps

D flux vs time: positive hump then negative hump

Question 288
Justify your answer to the previous question.

[3 marks]

Question 289
Which of the following best describes how the emf changes with time?

A   It rises from zero to a maximum and then drops back to zero.

B   It rises from zero to a maximum and then drops back to zero. It does this twice.

C   It rises from zero to a negative maximum, drops back to zero, increases to a maximum in the opposite direction, and then drops back to zero.

D   It rises from zero to a maximum and stabilises on that value.

Question 290
Justify your answer to the previous question.

[3 marks]

Question 291
If the change in field from t = 0 to t = 0.15 s (when the magnet is in the centre of the coil) is 0.05 T, the area of the coil is 0.2 m², the number of turns is 100, find the average emf induced during this time.

A one turn coil (area = 0.15 m²) of wire lies flat between the poles of an *electromagnet*, as shown. The field of the electromagnet changes as shown in the left hand graph following. The emf between A and B is graphed as a function of time, shown on the right below.

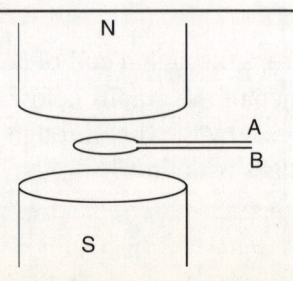

7 Generation principles

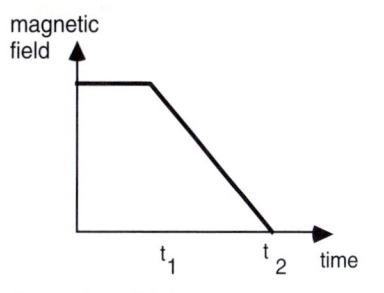

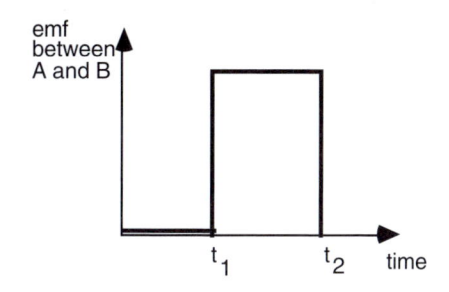

Question 292
Explain why the induced voltage varies with time as shown.

In a second experiment, the magnetic field is exactly *halved*, and the time interval between t_1 and t_2 is exactly *doubled*.

Question 293
Sketch the graph of the emf between *A* and *B* as a function of time.

Question 294
If the time interval between t_1 and t_2 is 25 ms and the initial value of the magnetic field is 200 mT, calculate the maximum value of the emf between *A* and *B*.

In a third experiment, the field in the electromagnet is again reduced to zero, reversed and increased again, as shown in the graph to the right.

Question 295
Sketch the shape of the emf induced between A and B in this case and explain your answer. [4 marks]

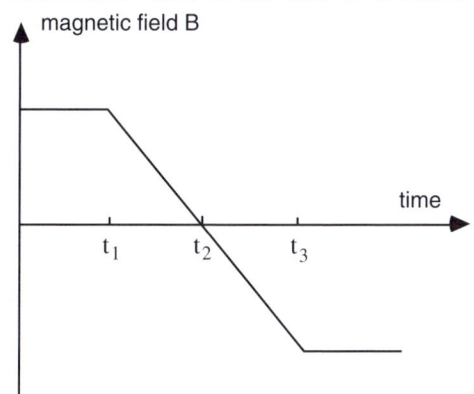

A falling weight (2 kg) is attached to a pulley on a small electric motor. The motor is a simple one, with permanent magnets supplying the magnetic field required. The terminals are connected to a circuit containing a switch and a small globe, as shown to the right.

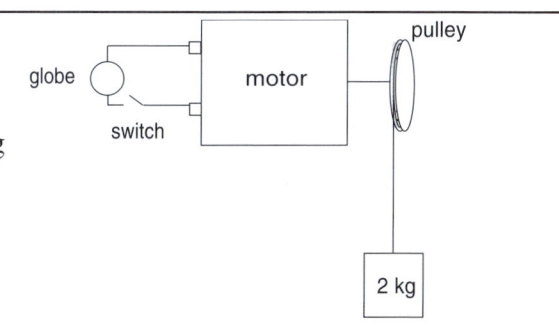

Question 296
Explain how it is that a *motor* can be used in this way to light a globe. Refer in your explanation to the basic physics principles involved.

[4 marks]

7 Generation principles

Question 297
When the switch is closed, the weight falls slower than when it is open. Explain this. Refer to the interaction between current and magnetic field. [4 marks]

Question 298
The lamp becomes brighter as the weight approaches the floor. Which one of the following best explains this?

A   The gravitational potential energy is reduced as the weight falls.

B   The friction in the motor bearings affects the movement of the weight in the early part of the fall.

C   The weight accelerates as it drops.

D   As with hydro-electric generation, the height of the drop affects the power output.

Question 299
The globe has a resistance of 4.0 ohm. At its brightest it draws a current of 0.6 A. What is the power output of the motor at this instant?

A student uses a single loop of wire and an electromagnet to investigate electromagnetic induction. The loop is placed between the pole pieces of the electromagnet, as shown, perpendicular to the magnetic field, and connected to an oscilloscope so that any induced voltage can be measured.

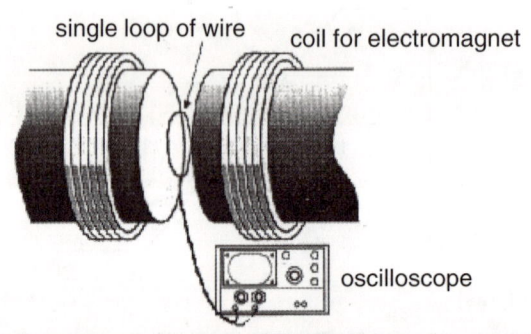

The current in the electromagnet coil is reduced so that the magnetic field, B, decreases to zero at a constant rate, as shown below. The induced voltage measured on the oscilloscope is also shown.

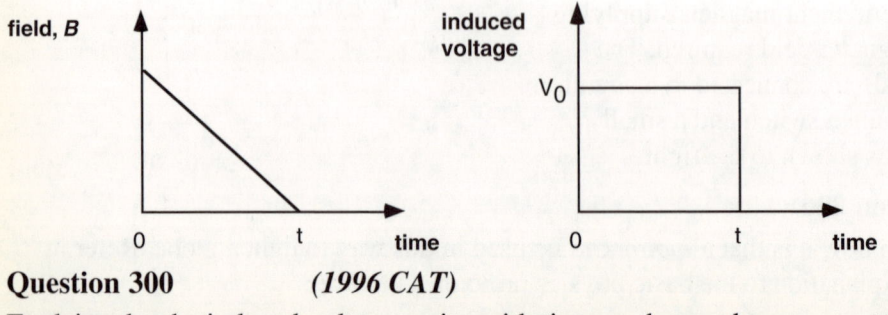

Question 300   (1996 CAT)
Explain why the induced voltage varies with time as shown above. [4 marks]

In a second experiment, the magnetic field is set to *twice* that used before. The field is now reduced to zero in a time t/2 (*half* the previous time).

Question 301 (1996 CAT)

Which one of the following figures (A–E) shown below best represents the variation of the induced voltage as a function of time?

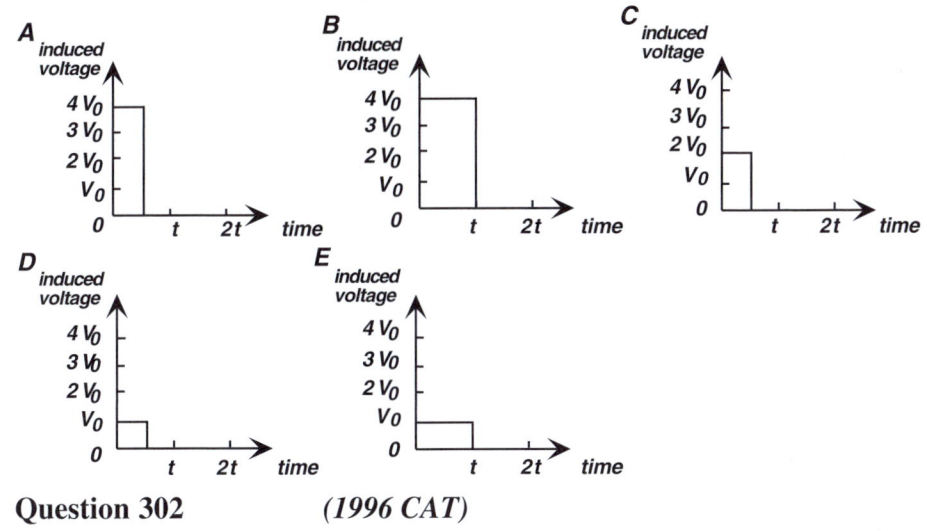

A — induced voltage: value $4V_0$ from 0 to t, then 0.

B — induced voltage: value $4V_0$ from 0 to 2t.

C — induced voltage: value $2V_0$ from 0 to t, then 0.

D — induced voltage: value $2V_0$ from 0 to 2t.

E — induced voltage: value V_0 from t to 2t.

Question 302 (1996 CAT)

In another experiment, the magnetic field has an initial value of 1.2 T, and is reduced to 0 at a constant rate in a time of 0.010 s. The magnitude of the induced voltage is 0.060 V. What is the area of the single loop of wire?

A student investigates electromagnetic induction using a single loop coil and an electromagnet. The loop is placed between the poles of the electromagnet, perpendicular to the magnetic field and connected to an oscilloscope so that any voltage induced in the loop can be measured.

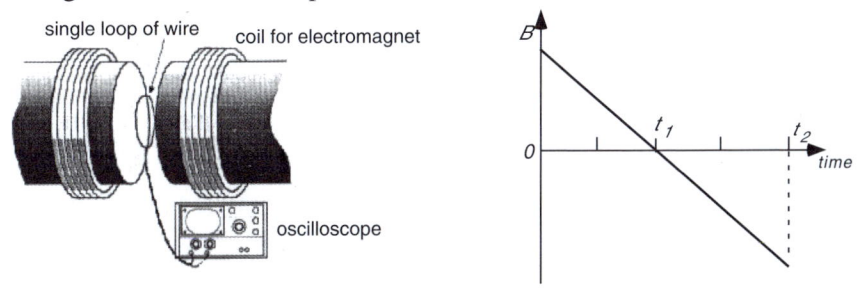

The current in the coils of the electromagnet is reduced to zero and then reversed so that the magnetic field, B, changes as shown above.

Question 303 (1997 CAT)

Which one of the graphs following best represents the induced voltage measured on the oscilloscope? Give reasons for your answer, referring to the diagram and graph above. [4 marks]

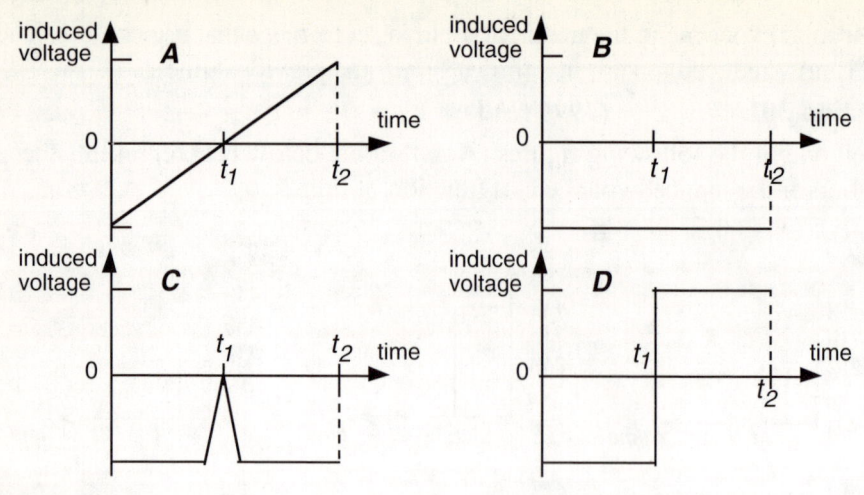

Question 304
When the current in the coils is a maximum value, the magnetic field between the poles has a value of 0.15T. The diameter of the single loop coil is 2.0 cm. At maximum current value, what is the magnetic flux through the coil?

Gary and Kate are investigating electromagnetic induction. They have a single wire loop of dimensions 0.030 m by 0.020 m, which is placed in a uniform magnetic field, of strength 0.12 T. The loop can be rotated by hand about an axis as shown in the diagram below. The ends of the loop slide within slip rings so that a measurement of the emf between the ends of the loop can be made between terminals A and B.

Question 305 *(1998 CAT)*
What is the value of the magnetic flux through the loop when it is oriented as shown above?

Question 306 *(1998 CAT)*
When the coil rotates at a constant rate of 5 Hz, the voltage between points A and B varies with time as shown in the graph below. Sketch the *voltage* variation with time if the rate of rotation is doubled.

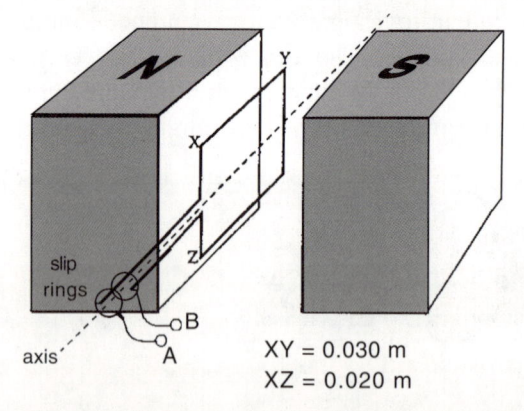

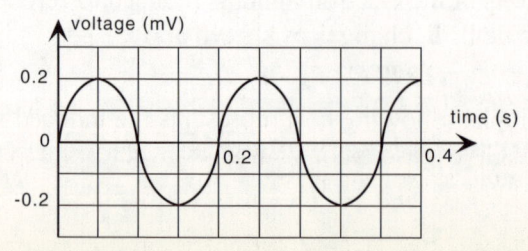

XY = 0.030 m
XZ = 0.020 m

7 Generation principles

Question 307

Sketch the variation of the *magnetic flux* for these conditions. Use the same time scale on your graph. You do not need to label the vertical axis with values.

[3 marks]

Sofia and Max are investigating electromagnetic induction using a square coil. They place the coil between the poles of a magnet as shown to the right. The sides of the coil are 0.020 m long. The uniform magnetic field between the poles is 5.0×10^{-2} T, and elsewhere in air it is assumed to be zero.

Question 308   (1999 CAT)

Calculate the magnetic flux through the coil when it is entirely within the magnetic field, as in the diagram to the right.

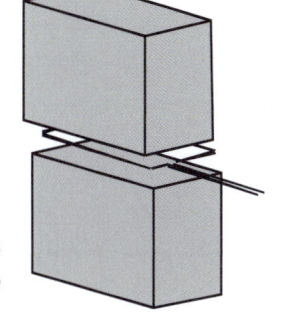

The coil is moved from the position shown above (the coil entirely within the magnetic field) to the position shown at the right (the coil entirely outside the magnetic field). This takes 0.040 s.

Question 309   (1999 CAT)

What is the average emf induced in the coil as it moves from the first position to the second?

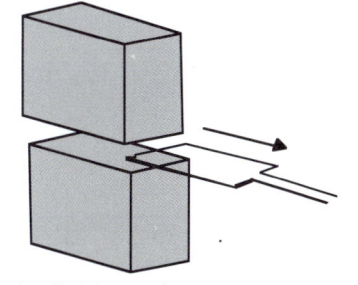

Sofia and Max now move the coil through the magnetic field, as shown.

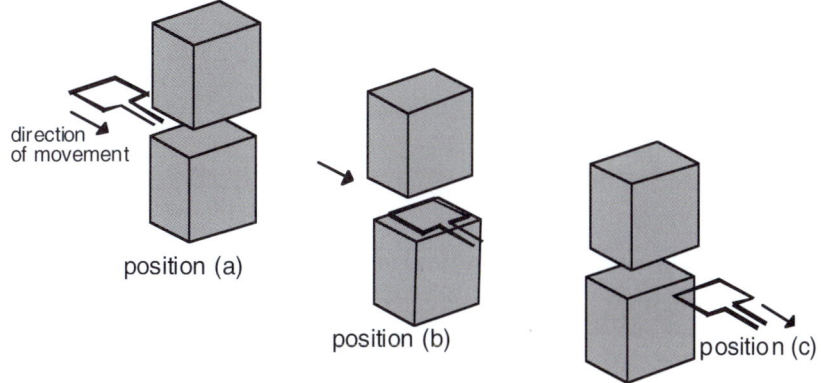

direction of movement

position (a)

position (b)

position (c)

The magnetic flux through the coil as it moves from (a) to (b) to (c) is graphed to the right.

Question 310   (1999 CAT)

Which of the following graphs best shows how the emf induced in the coil changes with time?

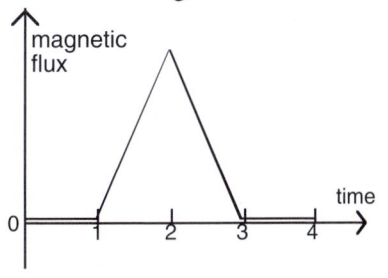

77

7 Generation principles

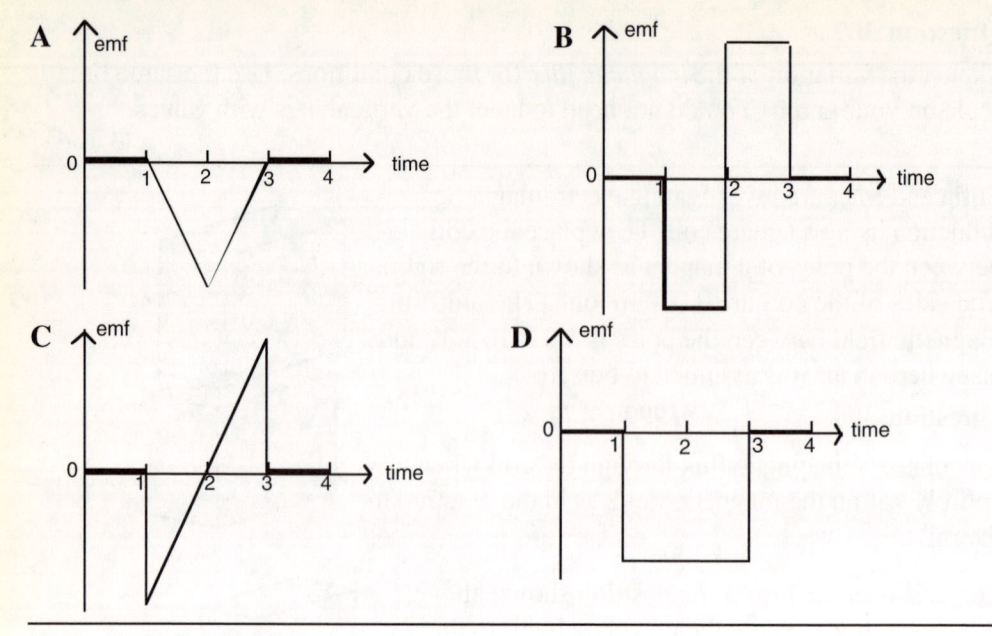

A emf vs time graph

B emf vs time graph

C emf vs time graph

D emf vs time graph

A simple DC motor has one coil and permanent magnets.

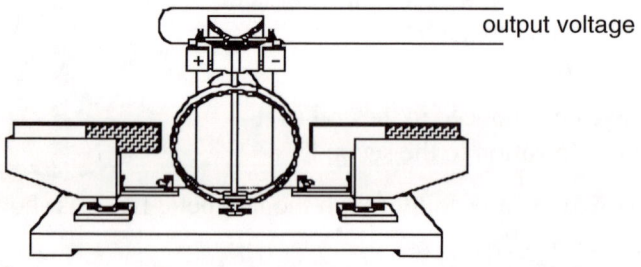

output voltage

Question 311

If spun by hand, it gives an output voltage. The reason for this is:
- **A** the flux through the coil changes.
- **B** the attraction of the two permanent magnets for each other forces the electrons around the coil.
- **C** this action will cause an equal and opposite reaction.
- **D** friction generates heat which accelerates the electrons.

A generator coil of area 0.2 m² has a magnet on the end of a lever which oscillates up and down through a coil. The sketch graph below shows how the magnetic flux through the coil changes with time as the magnet moves into and out of the coil.

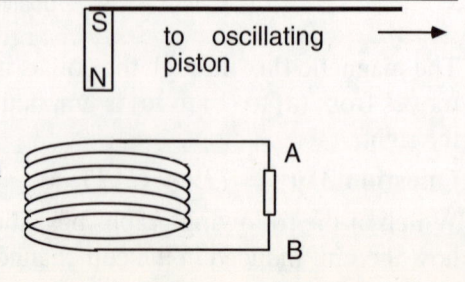

78

7 Generation principles

average flux through each turn

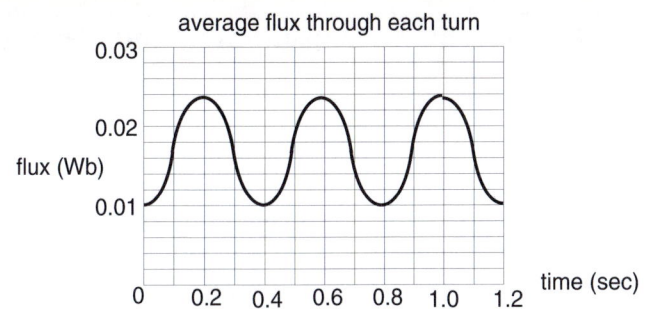

Question 312
Sketch the emf across the coil against time, between t = 0 and 0.4 s. [3 marks]

Question 313
Calculate the average emf between time t = 0.4 s and t = 0.6 s. (number of turns of the coil = 50).

Question 314
If the generator is attached to a load of resistance 4.5 Ω, calculate the average current that it will supply. The resistance of the coil is very low.

The on-off action of a lightning stroke produces an electromagnetic field surrounding the stroke. It is this field that causes the crackling in your radio or TV during a thunderstorm.

The magnetic field, **B**, produced by a lightning stroke varies with time as shown in the graph on the right.

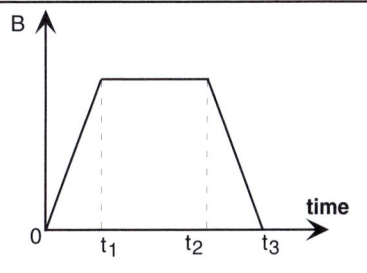

Question 315 *(2000 written examination)*
A small coil is placed perpendicular to the magnetic field, and the induced emf is monitored on an oscilloscope. Which *one* of the following graphs best shows the variation of the emf with time?

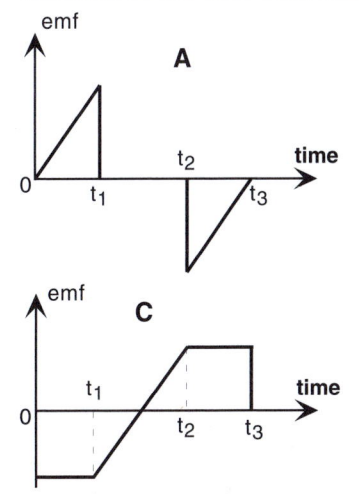

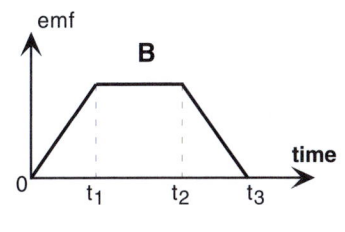

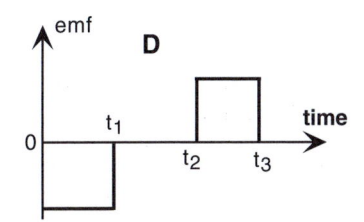

CHAPTER 8 — Transformers, transmission and consumption

Question 316
A step-up transformer is used to boost a voltage from 215 V to 240 V (both RMS). If the primary has 500 turns, calculate the number of turns needed for the secondary. Give your answer to the nearest turn.

Question 317
Lucy is making measurements on a transformer, as shown.

Coil	Voltage (RMS)	Current (RMS)
PRIMARY	240 V	1.2 A
SECONDARY	19.8 V	12 A

Which of the following is the most likely value for the ratio $\dfrac{primary\ turns}{secondary\ turns}$?

A 0.05 **B** 0.08 **C** 0.1 **D** 10 **E** 12 **F** 20

Question 318
Calculate the ratio: $\dfrac{power\ output}{power\ input}$

Question 319
Briefly explain why your answer to the question above is *not* equal to 1.

A transformer is used to provide a low voltage supply for decorative lighting on a tree in Syd's backyard. The circuit is represented below.

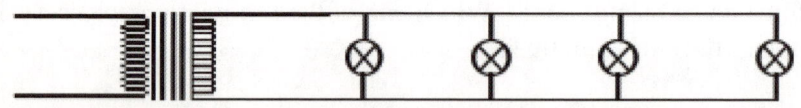

The input of the transformer is connected to 240 V AC. The globes in the circuit are designed to operate properly from 12 V AC.

Question 320
If the transformer can be considered ideal, calculate the ratio $\dfrac{primary\ turns}{secondary\ turns}$

Question 321
Explain why the input of the transformer *must* be AC rather than DC. Refer to the basic principle of physics involved.

[3 marks]

8 Transformers, transmission and consumption

Electric power is delivered to an overseas city with a range of different transmission voltages. All the voltages shown are RMS.

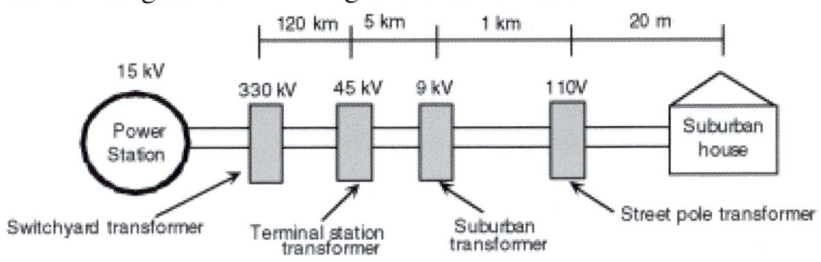

The power station generator produces energy at 15 kV RMS. In the adjacent switchyard this is stepped up to 330 kV RMS before transmission to the outer suburban area. The diagram shows the output voltages of the various transformers before delivery to each household at 110 V RMS.

Question 322
What is the value of the ratio

$$\frac{\text{number of turns on secondary of switchyard transformer}}{\text{number of turns on primary of switchyard transformer}} \ ?$$

At the primary of the street pole transformer, the current is 1.5 A, and the voltage is very close to 9 kV. The transformer can be assumed to have no significant power losses within it.

Question 323
What RMS current flows in the secondary of the street pole transformer?

Question 324
Although the switchyard transformer produces electricity at 330 kV, the input voltage to the next transformer is 325 kV. Explain why this occurs, quoting relevant formulas. Use the symbol R for the resistance of the transmission.

[3 marks]

Question 325
One form of household safety device, sometimes described as RCI, relies on detecting an *inequality* between the active and neutral currents in a circuit. Normally the active and neutral currents should be exactly equal. Explain why such an inequality might indicate a dangerous fault in a circuit. [4 marks]

One method of detecting any inequality is shown in the diagram below. The active and neutral wires both are wound around the transformer core in opposite directions to each other.

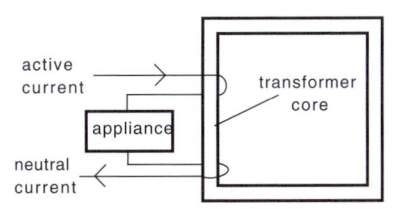

8 Transformers, transmission and consumption

Question 326
When the active and neutral currents are exactly equal, there should be no flux in the core of the transformer. Explain why.

Question 327
When the active and neutral currents are *not* equal, and they are both AC, which of the following best describes the situation?

A   There is no flux through the transformer core.

B   There is a steady flux through the transformer core.

C   There is a changing flux through the transformer core.

A country house receives its electricity some kilometres from the main supply cables. The cables involved have resistance which means that some voltage is 'lost' along them. The diagram below outlines the situation.

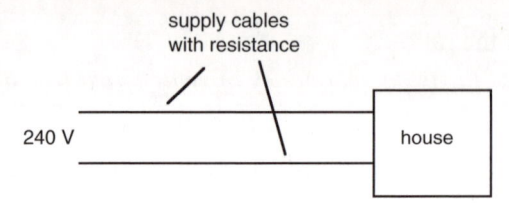

Question 328
The voltage received at the country house changes, depending on the current being drawn by the house. Explain why this occurs. [4 marks]

Question 329
On one occasion, the voltage received at the house is measured as 225 V. The current being drawn at the time is measured as 45 A. Calculate the resistance of the supply cables.

Question 330
John's average daily energy use from July – September 1995 was 32 kW h. Calculate the energy, in joule, he used over this period.

A substation (S) in the Melbourne metropolitan electricity distribution system is 80 km from the terminal station (T) in the Latrobe valley. The power delivered along a particular cable is 10 MW, at a voltage of 66 kV. The voltage drops from 66 kV at T to 65.8 kV at S.

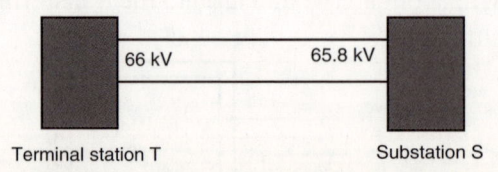

Question 331
Explain why this voltage drop has occurred.

82

8 Transformers, transmission and consumption

Question 332
Which one or more of the following is/are associated with this drop?
A   There is less power delivered to S than leaves T.
B   There is as much power delivered to S as leaves T.
C   The current at S in the cable is less than that at T.
D   The current at S in the cable is the same as leaves T.

Question 333
Calculate the power loss along the cable.

Question 334
Calculate the resistance of this cable.

Question 335
The cable is replaced by cheaper cable with a resistance of 0.24 Ω km^{-1}. What is the power loss now? Show your working. [3 marks]

You leave for school at 8:00 am in the morning but leave your 60 W bedside lamp on. You return home at 4:30 pm. The domestic tariff for electricity is 11.7 cents kW h^{-1}. Give your answer to the nearest cent.

Question 336
What is the value of the electrical energy used?

A farmer's 10 kW generator produces power 200 V. Power is transmitted over a 5 km distance without transformers. It arrives at the load at 170 V.

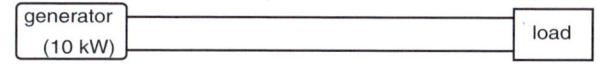

Question 337
What current is flowing in the wires?

Question 338
What is the total resistance of the transmission wires?

Question 339
How much power has been transformed to heat energy? Show working. [3 marks]

The farmer attempts to reduce power losses involved in the transmission by installing some transformers. They have a turns ratio of 1:20.

Question 340
Draw a sketch showing how they should be connected into the circuit.

Question 341
Calculate the current flowing in the wires with the transformers in place.

Question 342
Recalculate the voltage arriving at the load.

Question 343
Recalculate the power loss.

8 Transformers, transmission and consumption

Question 344
What is the ratio:
$$\frac{power\ loss\ with\ transformers}{power\ loss\ without\ transformers}\ ?$$

A 6 V supply for an electric light in a distant backyard shed is supplied as shown in the diagram. A long piece of 2-core flex is used to connect the shed light to the output of a 6 V transformer in the house.

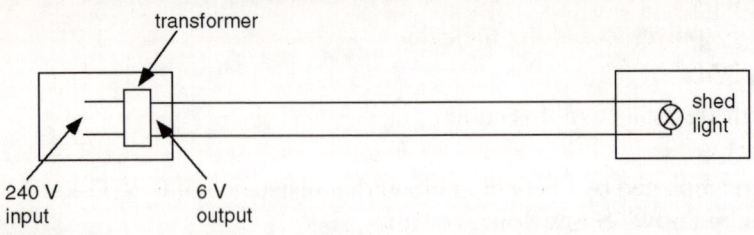

Question 345
If the transformer input coil contains 960 turns, how many turns are there on the output coil?

Question 346
When the shed light is turned on it draws a current of 3.5 A RMS. Which of the following best describes the current in the shed light?

A	A steady current of 3.5 A.
B	A varying voltage with a peak value of +3.5 A.
C	A varying voltage (peak value ~ 5 A) but an average of 0 A.
D	A varying voltage (peak value ~ 5 A) but an average of 3.5 A.

Question 347
The wires out to the shed have some resistance, with the result that the shed light operates off *less* than 6.0 V. The voltage at the light is 5.0 V. Calculate the total resistance of the wires leading out to the shed.

Question 348
Calculate the power dissipated in the wires leading out to the shed when the shed light is switched on.

A small industrial motor needs *at least* 15 kW of electric power. It has a working resistance of 0.80 Ω.

Question 349
Which of the following values of supply voltage would be suitable?

A	12 V	B	24 V	C	50 V
D	100 V	E	120 V	F	150 V

Thomasina the steam engine has been converted into an electric locomotive. Her electric engines need 200 A at 100 V (both RMS) to operate normally. Electric power is only available from the overhead cables at 1000 V, so she has a transformer installed on board.

Question 350
What is the current (RMS) she will need to draw from the cables if she is to operate normally? (You can assume that there are no losses in the action of the transformer.)

Question 351
A section of the load curve of the electricity supply for a small country town is shown. All values are RMS. What is the energy consumption (in kW h) between 1:00 pm and 4:00 pm? Show your working.

[3 marks]

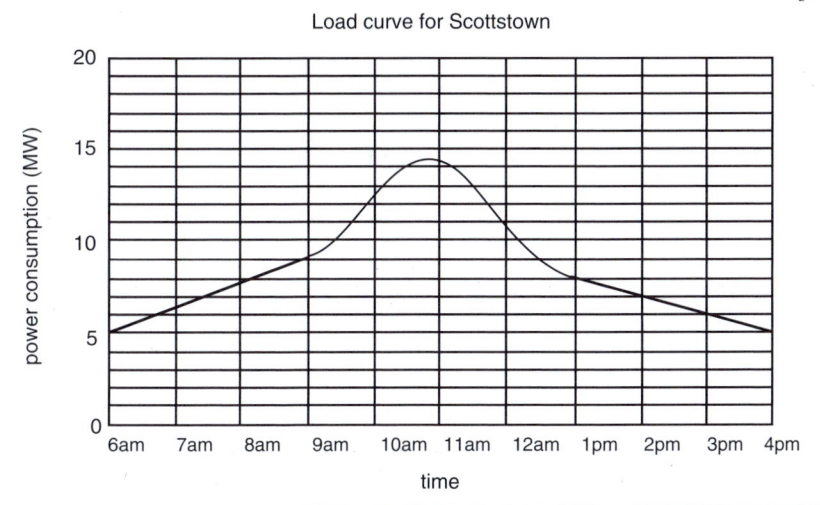

Load curve for Scottstown

A farmer installs a low-voltage light at the front gate of a farm which is 200 m from the farmhouse. A transformer, situated in the house, is used to convert the 240 V RMS supply to 6.0 V RMS, and a long 2-wire cable is connected to the light as shown below in Figure 2.

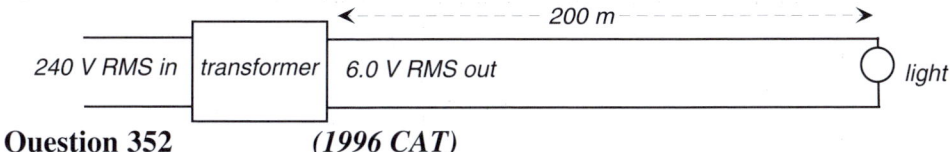

240 V RMS in | transformer | 6.0 V RMS out — 200 m — light

Question 352   (1996 CAT)
Write down the value of the ratio

$$\frac{\text{number of turns on the transformer secondary coil}}{\text{number of turns on the transformer primary coil}}$$

Question 353   (1996 CAT)
When the light is on it draws a current of 4.0 A RMS. The total resistance of the wires from the transformer to the light and back is 0.5 Ω. Calculate the power dissipated in the light globe. Show your working.

[4 marks]

8 Transformers, transmission and consumption

Question 354   *(1996 CAT)*

Which one of the graphs (A – D) following best represents the variation with time of the current in the light circuit?

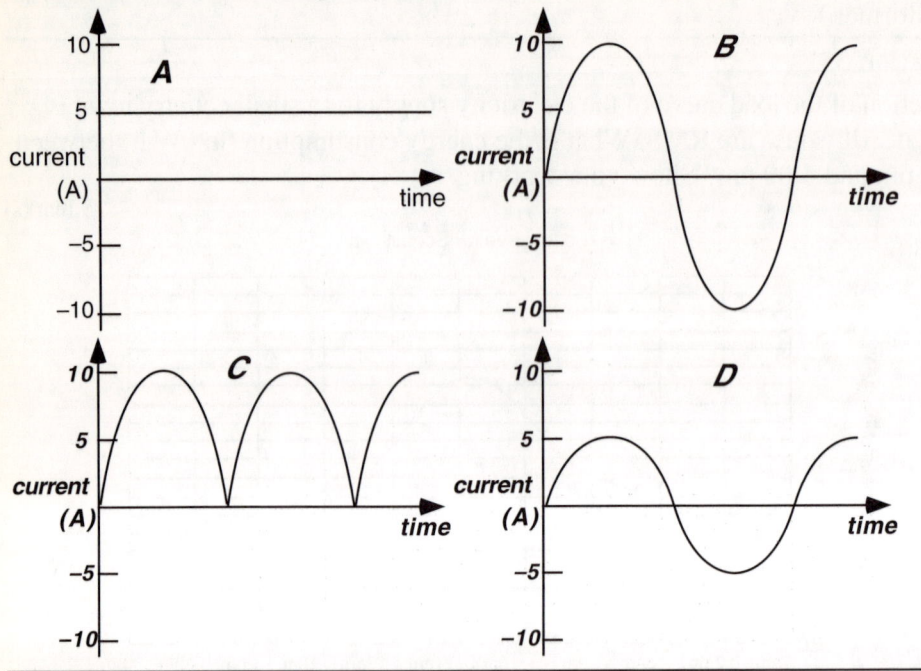

A generator at a ski resort supplies electricity to a small ski tow, with two motors, A and B, and lighting. The tow is distant from the generator. Wiring is shown. The RMS voltage at the generator is 240 V. The RMS voltage at the tow is 200 V. When operating, the RMS current delivered to motor A is 12 A, to motor B is 9 A, and to the lighting is 4 A.

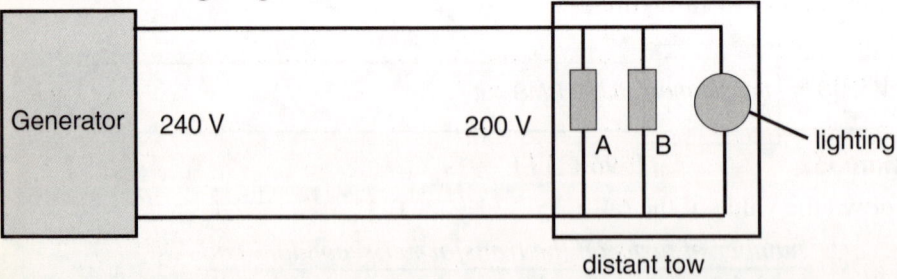

Question 355
How much power is supplied when all three appliances are operating?

Question 356
How much power is lost in the electric cables to the tow? Show working.
[3 marks]

Question 357
What is the effective resistance of the cables?

8 Transformers, transmission and consumption

An electrician has imported an electric light globe designed to operate at 110 V (RMS). At this voltage it has a resistance of 55 ohm. The globe's power rating is 220 W. Two methods are considered that will allow the light globe to be used with a mains supply of 240 V (RMS). Method 1 uses a resistor of 65 Ω placed as shown below, so that when used with an RMS supply voltage of 240 V, the RMS voltage across the light globe is 110 V.

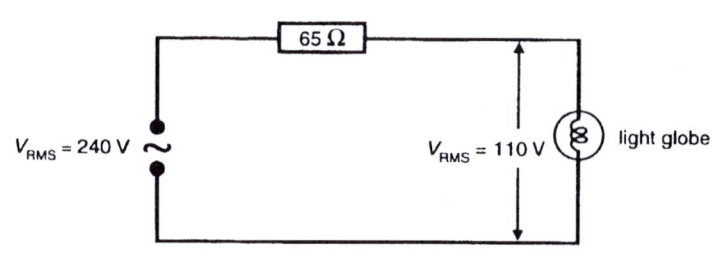

Method 2 uses a transformer to convert the 240 V to 110 V (RMS). Assume the transformer is ideal.

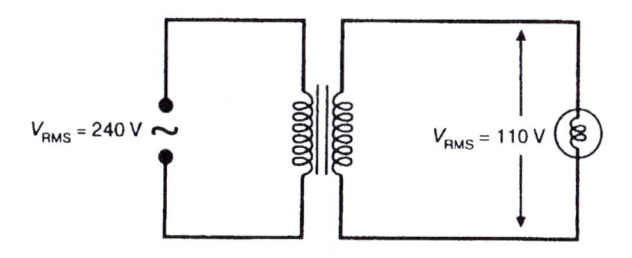

Question 358   *(1997 CAT)*
There are 1440 turns on the primary coil of the transformer. How many turns are on the secondary coil of the transformer?

Question 359   *(1997 CAT)*
What is the RMS current in the primary coil of the transformer?

Question 360   *(1997 CAT)*
What power is supplied from the mains for the two different methods?

Fred's average daily usage of electrical energy for a year is shown below.

8 Transformers, transmission and consumption

Question 361   (1997 CAT)
Estimate the total electrical energy (in joule) used by the household in the three months of June, July and August 1996.

Question 362   (1997 CAT)
Supply costs are 12 c/kW h. What does a 2.4 kW heater cost for 8 hours?

Question 363
The increase from 10 kW h in summer to 30 kW h in winter is due to the 4.0 kW electric central heating system. How many hours each day during mid-winter, on average, is the electric central heating system operating?

A system for delivering power to an isolated pump from the 240 V generator that supplies a cattle station, is shown below.

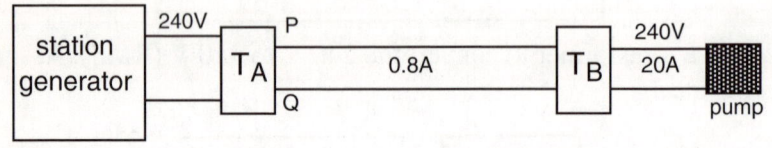

```
[station generator | 240V ] —— P ——[ T_A ]—— 0.8A ——[ T_B ]—— 240V 20A ——[pump]
                                      Q
```

The two transformers T_A and T_B were installed when it was realised too much power was being lost in transmission. Assume no power is lost in the transformers.

Question 364
Explain why T_A and T_B increase the fraction of generator output power transferred to the pump. Your answer should clearly show the nature (step-up or step-down) of the transformers, and the physics involved.

Question 365
The RMS current in the transmission wires is 0.8A. What is the output voltage V_{PQ} from T_A?

Question 366
The secondary coil of T_A consists of 100 turns of wire. How many turns of wire are in the primary coil of T_A?

Question 367
If the transmission wires have a combined resistance of 4.0 Ω, what power is dissipated in these?

Before the transformers, wires connected the generator and pump directly. The pump, as in the current arrangement, drew 10 A of current.

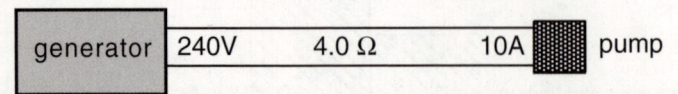

```
[generator | 240V] —— 4.0 Ω —— 10A ——[pump]
```

Question 368
Under this arrangement, how much power was dissipated in the cables?

8 Transformers, transmission and consumption

The graph below shows the average daily usage of electricity in a house. The vertical scale is in kilowatt-hour (kW h).

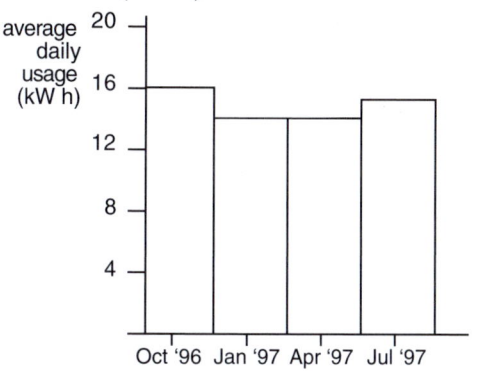

Question 369   *(1998 CAT)*

The unit kW h on the graph is not in common use in academic physics. Which of the following units could be used (with a different scale) on the vertical axis instead of kW h.

A J   **B** J s   **C** J s^{-1} **D** W

Question 370   *(1998 CAT)*

The average daily usage for the 3 month period September–October–November 1996 is 16 kW h. All this electricity was used to heat the house with an electric heater with a power consumption of 2000 W. How many hours each day, on average, was the heater used for?

A 4.0-kW electric generator is used to supply power to an electric motor some distance away. The RMS voltage at the output of the generator is 400 V.

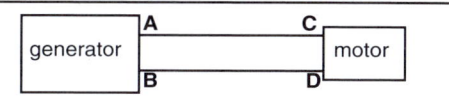

The RMS current supplied to the motor is 10 A. The total resistance of the two cables, AC and BD, is 3.0 Ω.

Question 371   *(1998 CAT)*

How much electric power is dissipated in the cables?

An engineer suggests that by connecting a transformer (T_1) at the generator output, and another (T_2) at the motor input, a larger fraction of the generated power could be transferred to the motor, as shown.

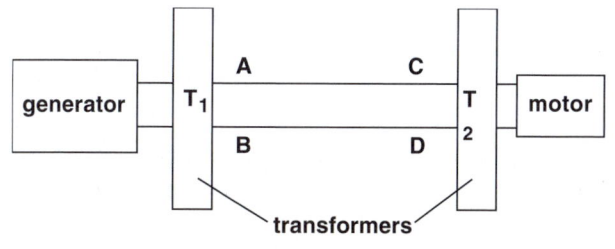

8 Transformers, transmission and consumption

Question 372   *(1998 CAT)*

Explain why this procedure increases the power transferred to the motor. Your answer should clearly show the physics involved. Indicate what type of transformer (step-up or step-down) should be used in each location.   [4 marks]

Andreas is having trouble with the reading lamp on his desk. Globes rated at 60 W and designed for an RMS voltage of 240 V keep burning out. He seeks help from his friend Emma who is a qualified electrician. They measure the RMS supply voltage and find it to be 264 V.

Question 373   *(1999 CAT)*

What would the RMS current in one of these 60 W globes be if it was connected to an RMS voltage of 240 V?

Emma and Andreas have different solutions to the problem of the globes burning out. Emma suggests to solve the problem by stepping down the supply voltage from 264 V to 240 V, using a transformer with 1000 turns in the primary coil as shown below.

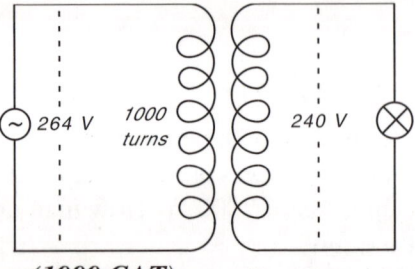

Question 374   *(1999 CAT)*

How many turns would there be in the secondary coil, assuming the transformer is ideal?

Andreas suggests the problem could be solved by connecting an appropriate second globe (globe X below) in series with the 60 W globe.

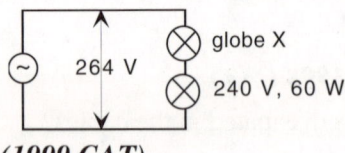

Question 375   *(1999 CAT)*

What would be the power used in globe X?   [4 marks]

A power station in the Latrobe Valley generates electric power at an RMS voltage of 20 kV (20 000 V). The switchyard transformer steps up the voltage to an RMS value of 500 kV (500 000 V) for transmission to other parts of Victoria. On the outskirts of large towns a local-area transformer steps down the voltage for local transmission. The circuit below shows how the power station and the two transformers are connected. The RMS current in the secondary of the switchyard transformer is 300 A. Assume the two transformers are ideal.

8 Transformers, transmission and consumption

current = 300 A

20 kV — 500 kV — transmission lines (total R = 8.0 Ω)

switchyard transformer — local area transformer

Question 376 *(1999 CAT)*
What is the RMS current in the primary of the switchyard transformer? The total resistance of the transmission lines is 8.0 Ω.

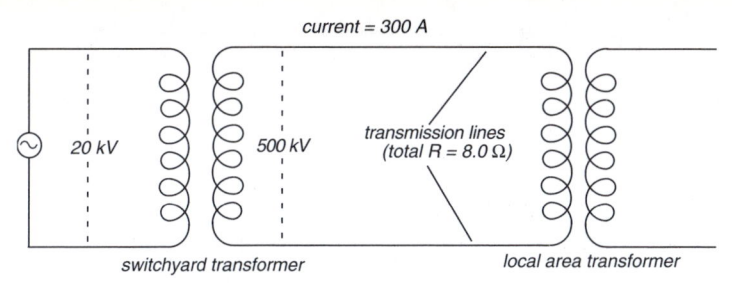

Question 377 *(1999 CAT)*
How much power is lost in the transmission lines?

Question 378 *(1999 CAT)*
What is the RMS voltage across the primary of the local-area transformer? Give your answer to three significant figures and make sure you show your working clearly.

A step-down transformer has a turns ratio of 20:1. It is designed to operate from the mains supply (240 V). It is used to produce higher currents than would normally be available from the mains supply.

Question 379
The mains can supply 10 A at 240 V. What is the maximum current that the step-down transformer could supply from its secondary coil?

Question 380
It is generally undesirable to connect the secondary of such a transformer to high resistance loads if large currents are required. Explain why. [4 marks]

Question 381 *(2000 written examination)*
Joe and Jan are installing two low-voltage lights in their garden. The lights are supplied from a transformer that has an output RMS voltage of 12 V, and is connected to the 240 V household supply.

What is the value of the ratio

$$\frac{\text{number of turns on the primary coil}}{\text{number of turns on the secondary coil}}?$$

Question 382 *(2000 written examination)*
Each light is designed to operate at an RMS voltage of 12 V, and has a resistance of 18 Ω, which does not depend on temperature.

What is the power dissipated in such a light when operated at an RMS voltage of 12 V?

Joe and Jan now connect light 1 to the transformer using two wires, each 16.0 m long. Each wire has a resistance of 0.050 Ω m⁻¹.

91

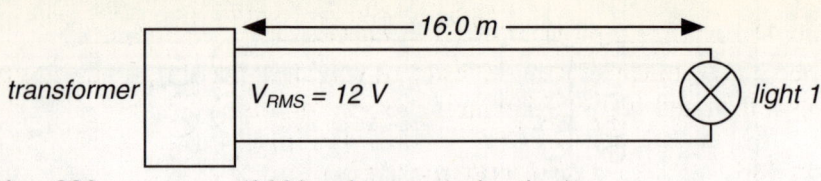

8 Transformers, transmission and consumption

Question 383   *(2000 written examination)*
What is the RMS voltage across light 1? Show your working.
[4 marks]

They now connect light 2 directly across the secondary of the transformer as shown in the diagram below.

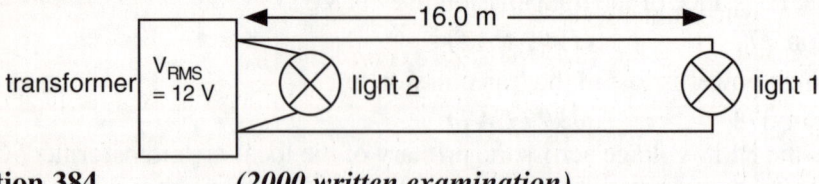

Question 384   *(2000 written examination)*
Joe and Jan had thought that with the circuit above, the two lights would be equally bright. In fact, light 2 is brighter than light 1. Explain why this is so. (Your answer should include the values
[4 marks]

Question 385   *(2000 written examination)*
Light 1 is now unplugged. Which of the statements below best describes the change in the brightness of light 2?
A   light 2 gets less bright
B   light 2 gets brighter
C   light 2 gets much brighter and burns out
D   light 2 does not change in brightness

It is common practice for the wires in the cables associated with garden lights to carry only low voltages (often 12 V_{RMS}). However it is more efficient to use 240 V globes for the lights. In order to achieve this, the circuit shown below can be used. At the 240 V supply, the voltage is stepped down using a 240 V to 12 V transformer, and at the light it is stepped up using a 12 V to 240 V transformer. The wires joining the two transformers are each many metres long. The transformers can be assumed to be ideal.

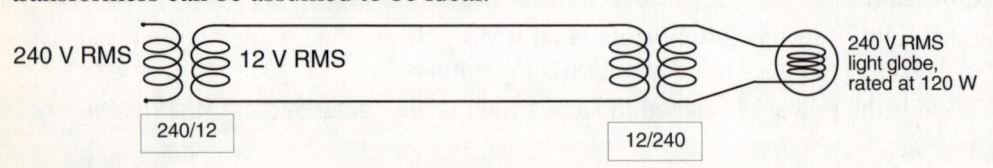

Question 386   *(2001 written examination)*
The light globe is rated at 120 W when connected to a 240 V_{RMS} supply. What current should flow through it under this condition?

Question 387 (2001 written examination)

When the system shown above was tested, it was clear that the globe was not operating at the rated 120 W. Explain the reason for this.

[3 marks]

Question 388 (2001 written examination)

When the garden light operates, the voltage across the input to the transformer that supplies the globe is 10 V_{RMS}. What is the voltage across the globe?

Question 389 (2001 written examination)

Under these conditions the current flowing through the long wires is 8.3 A. What current is flowing in the globe?

Question 390 (2001 written examination)

What is the total resistance of the two wires? (Remember that the transformers are ideal.)

Question 391 (2001 written examination)

Which *one or more* of the following changes would increase the voltage across the globe?

A   use wires of higher resistance
B   use wires of lower resistance
C   use transformers with ratios of 240:24 and 24:240
D   use transformers with ratios of 240:6 and 6:240

CHAPTER 9   **Electronics basics**

An electronics engineer is observing electrical signals inside the beam transporter on the 'Enterprise'. She is using a CRO and it is correctly adjusted. The first signal she observes is shown on the right.

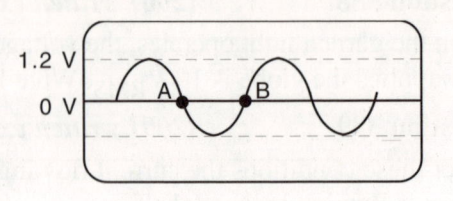

Question 392
She measures the distance between A and B to be 3.60 ms. What is the frequency of the signal?

Question 393
What is the RMS value of the voltage displayed on the CRO screen?

At another part of the apparatus, the following signal appears.

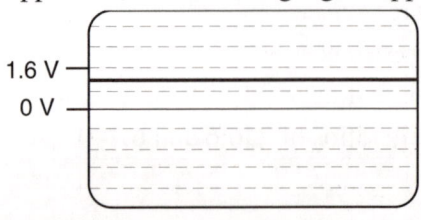

Question 394
Estimate the value of DC that this signal is closest to.

At yet another part of the apparatus the signal shown below occurs.

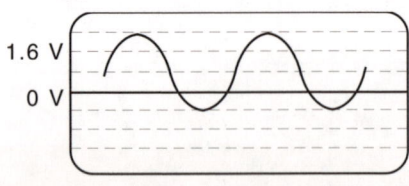

Question 395
This signal is best described by which of the following?

A   An AC signal of 2.4 V peak.
B   A signal with an average DC value of around 1.2 V.
C   A combination of DC and AC.
D   An AC signal of RMS value of 1.0 V.

9 Electronics basics

Question 396 to 400
Find V out for these circuits. All the resistors have the same value.

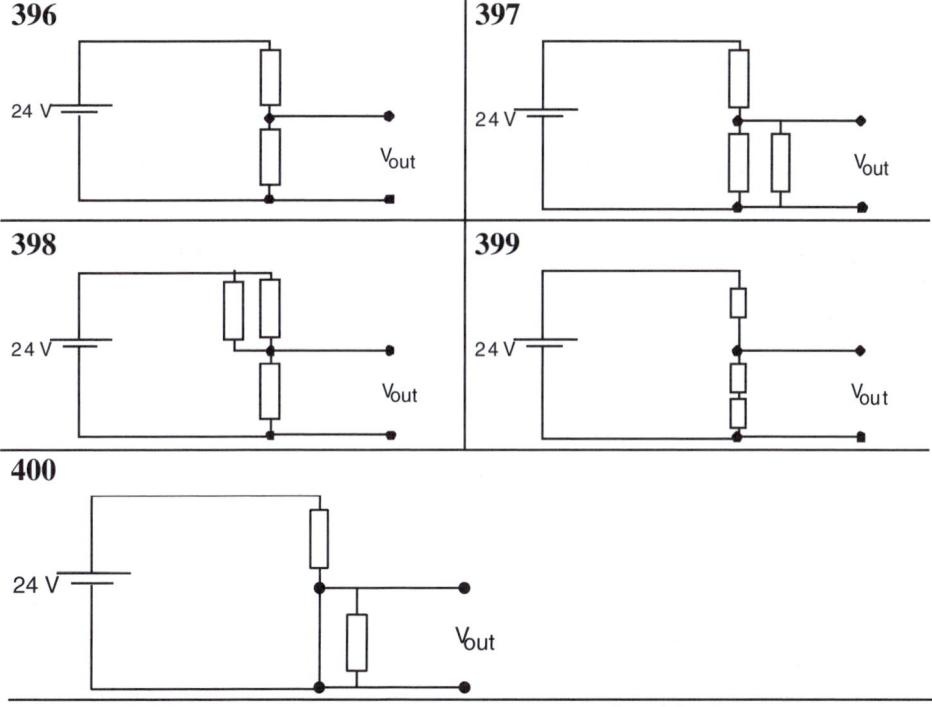

396

397

398

399

400

A temperature warning system is shown to the right. It is designed to sound the buzzer when the temperature falls past a certain 'danger' level. Initially point A is at a voltage of 1.2 V.

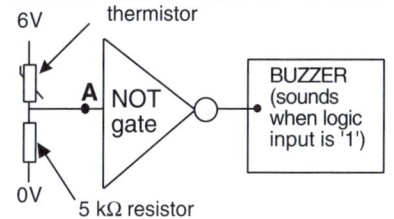

Question 401
What is the initial resistance of the thermistor?

Question 402
When the temperature falls just past the danger level, the output of the NOT gate changes from logic 0 to 1. The switching voltage of the NOT gate is 0.8 V. What is the resistance of the thermistor at this point?

A light dependent resistor is connected into a voltage divider as shown.

Question 403
What is the voltage at the point A?

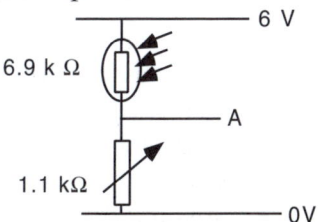

9 Electronics basics

Question 404
A signal generator provides an 250 Hz RMS voltage of 5.0 V. Assuming that the signal is sinusoidal, what is the peak value of the voltage?

Question 405
What is the *period* of this signal?

Question 406
This signal generator is connected in *series* with a DC cell with a voltage of 1.5 V, and a correctly adjusted CRO.

On the grid below, sketch correct values on the axes (to one significant figure).

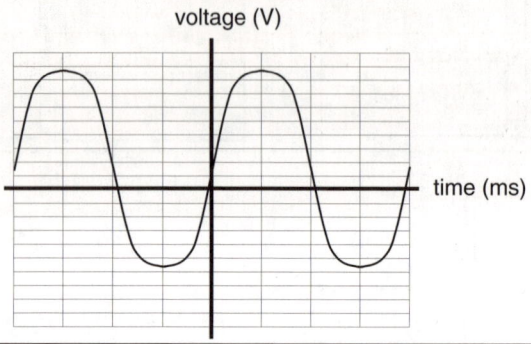

A thermistor's resistance is shown in the graph following.

Question 407
What is the thermistor's resistance at 25 °C?

The thermistor can be used as part of a temperature warning system, in a voltage divider, as shown to the right. The thermistor is used to monitor the temperature of a greenhouse. At one stage, the voltage across it is 3 V and the current through it is 5.0 mA.

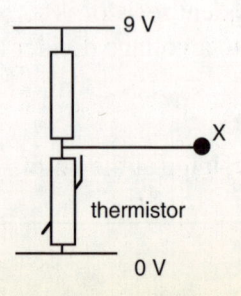

Question 408
What is the temperature of the greenhouse?

Question 409
As the temperature increases, the voltage at the point X changes. Will it increase or decrease? Give reasons from the information above.

[3 marks]

Question 410
At one time, the voltage at the point X is 4.2 V. The positions of the resistor and the thermistor are then reversed. What will the voltage at the point X be now?

Solar cells are devices that provide electrical energy. Their characteristics can be graphed as shown below.

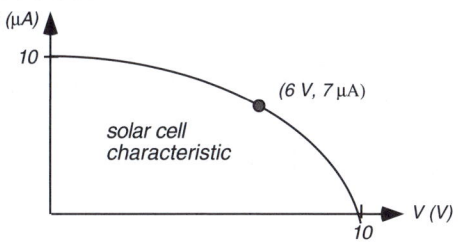

Question 411
What is the maximum voltage and current that the solar cell can provide?

Question 412
Calculate the power output of the solar cell at the point (6 V, 7 µA).

Question 413
Explain why power output of the solar cell can never be as large as the product of the maximum voltage and maximum current.

[4 marks]

The graph below shows the resistance of an LDR as a function of light illumination measured in lux. Note the logarithmic scales. The circuit below may be used to calculate the light illumination by measuring V_{out}.

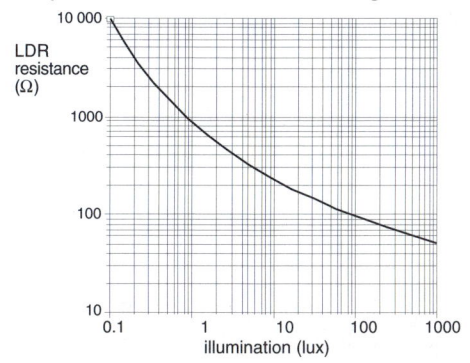

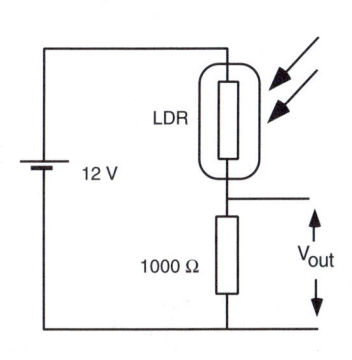

9 Electronics basics

Question 414 *(1996 CAT)*
What is the voltage V_{out} when the LDR has a resistance of 1400 Ω?

Question 415 *(1996 CAT)*
Explain what happens to the output voltage V_{out} as the light illumination on the LDR increases.

[4 marks]

Question 416 *(1996 CAT)*
When V_{out} is 6 V, what is the light illumination, in lux, on the LDR?

A circuit contains a sinusoidal generator, a switch and five identical resistors, each 20Ω, as shown.

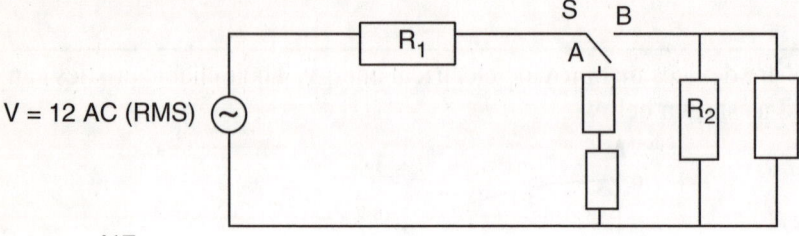

Question 417
With the switch set on position A, what is the current in resistor R_1?

Question 418
With the switch set on position B, what is the voltage across resistor R_2?

An electrocardiograph (ECG) is used to monitor heart activity. The ECG signal from a patient is displayed on a CRO, as shown below.

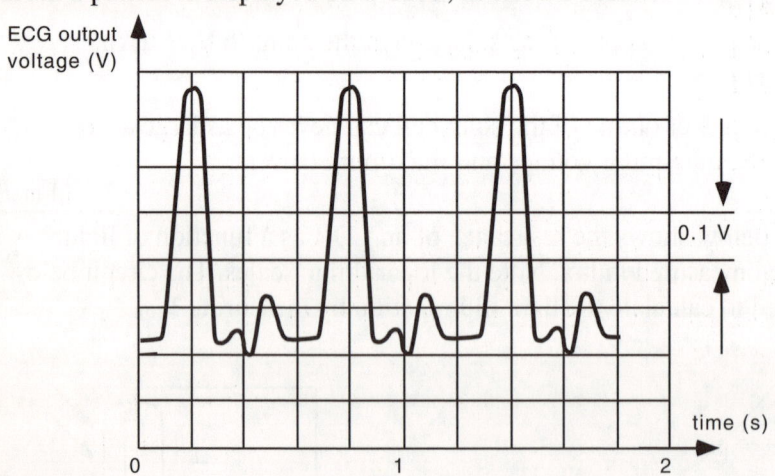

Question 419 *(1997 CAT)*
State the period of the voltage signal graphed above.

Question 420 *(1997 CAT)*
Use information from your answer to the question above to determine the heart rate of the patient in beats per minute. Show your working.

Question 421 (1997 CAT)
A circuit comprising a sinusoidal voltage generator, a switch and five resistors is shown below. Each resistor has a resistance of 100 ohms. The voltage of the generator is 15 V. What is this as a peak-to-peak voltage?

Question 422 (1997 CAT)
With the switch set to position A, what is the current in resistor R_1? Express your answer as an RMS current in mA.

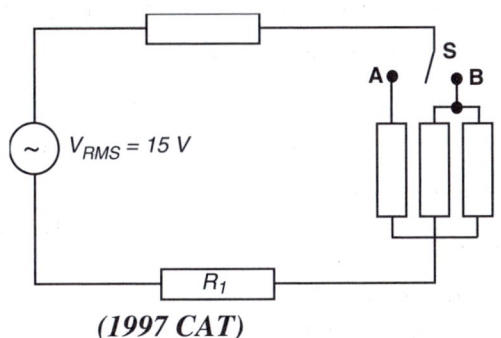

Question 423 (1997 CAT)
With the switch set to position B, what is the voltage across the resistor R_1? Express your answer as an RMS voltage.

The graph below shows the current-voltage characteristics of a diode. When a voltage of 0.8 V is placed across the diode, the current measured is equal to that obtained when the diode is replaced by a 10 ohm resistor.

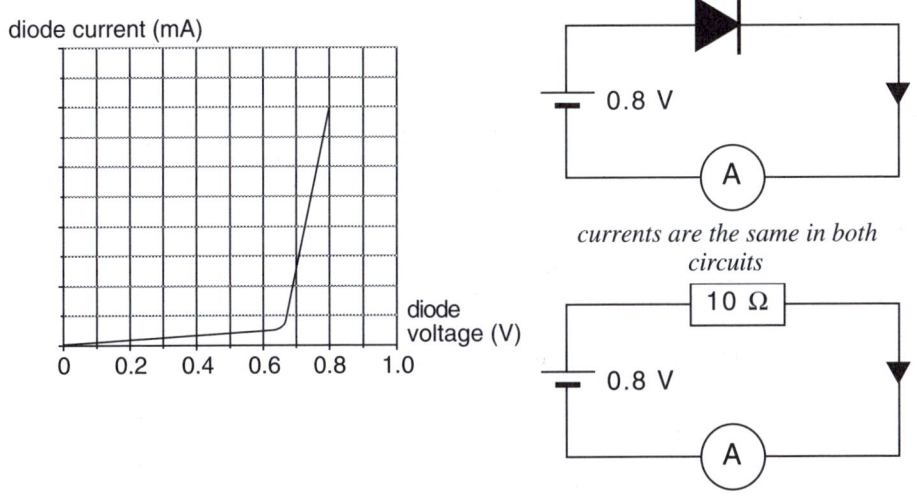

Question 424 (1997 CAT)
Using the information above, calculate the correct numbers for the current axis scale of the graph.

The same diode is placed in a circuit shown following. The battery voltage (E) can be varied. When $E_1 = 0.8$ V the current is I_1.

9 Electronics basics

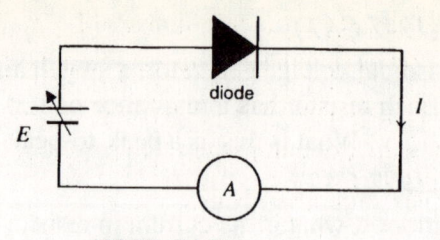

Question 425 *(1997 CAT)*

The battery voltage is now halved, that is $E_2 = 0.4$ V. Which one of the following expressions best describes the resulting current I_2?

A $I_2 = I_1$ B $I_2 > \frac{1}{2} I_1$ C $I_2 = 2I_1$

D $I_2 < \frac{1}{2} I_1$ E $I_2 = \frac{1}{2} I_1$ F $I_2 = 0$

Question 426

The circuit to the right shows a 12 V battery connected to two lights in series. L_1 has a resistance of 15 Ω, and in this circuit has 3 V across its terminals. What is the resistance of L_2? Show your working. [3 marks]

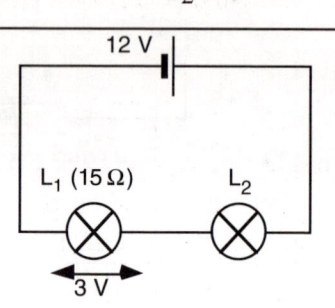

A sinusoidal voltage is shown.

Question 427

What is its RMS value?

Question 428

What is its frequency?

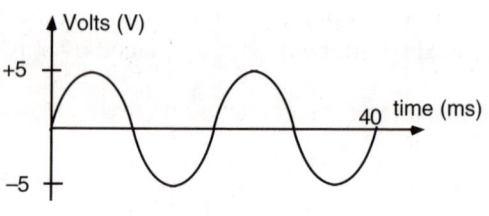

The diagram shows two electric lights connected in series to a 9.0 V battery. The lights have resistances R and 2R as indicated, when the current in the circuit is 300 mA.

Question 429 *(1998 CAT)*

Show that the voltages across the lights L_1, and L_2 are 3.0 V and 6.0 V respectively.

Question 430 *(1998 CAT)*

Determine the value of R in ohms.

Question 431 *(1998 CAT)*

The two electric lights are now connected in parallel as shown. Note that a 3.0 V battery is now used.

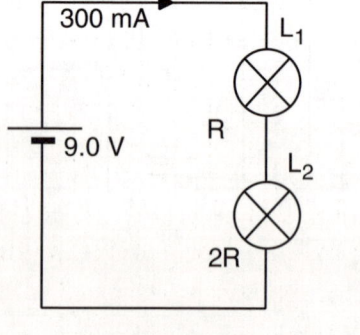

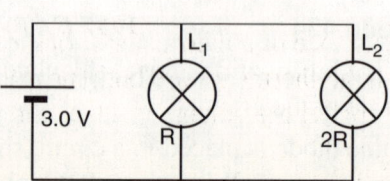

100

Which of the following best describes the brightness of L_1 in the two circuits? Support your answer with power calculations in both circuits.

A L_1 is brighter in the upper circuit than in the lower circuit.
B L_1 is less bright in the upper circuit than in the lower circuit.
C L_1 is the same brightness in both circuits.

[4 marks]

The signal generator shown on the diagram below provides electric power to a loudspeaker that has a resistance represented by $R_L = 10$ ohm. The switch S changes the power dissipated in the loudspeaker by selecting one of three resistors, $R_1 = 40$ ohm, $R_2 = 20$ ohm and $R_3 = 10$ ohm. The output RMS voltage of the signal generator is 0.50 V at a frequency of 1000 Hz.

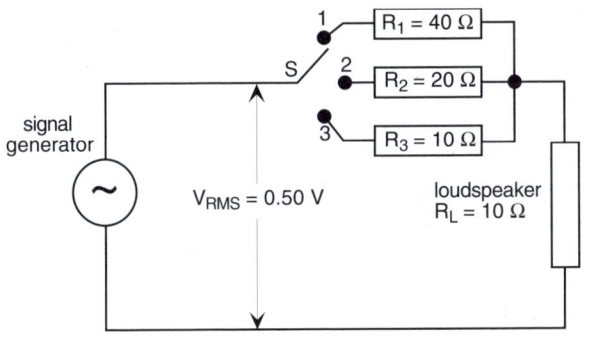

Question 432   *(2001 written examination)*
What is the RMS voltage across the loudspeaker if the switch is at position 1 (this selects $R_1 = 40$ ohm)?

Question 433   *(2001 written examination)*
The switch is now placed at position 2 (this selects $R_2 = 20$ ohm). Which one of the choices below gives the correct value of the ratio

$$\frac{\text{RMS current through resistor } R_2}{\text{RMS current through } R_L} \; ?$$

A   0.50   B   0.71   C   1.0   D   1.4   E   2.0

Question 434   *(2001 written examination)*
The switch is now placed at position 3 (this selects $R_3 = 10$ ohm). How much electrical power is dissipated in the loudspeaker now? (Give your answer in milliwatt.)

The resistance of a thermistor varies as shown. The voltage across and current through the thermistor are both measured. When the thermistor is placed in a cup of coffee, the voltage is 4.0 V and the current is 10 mA.

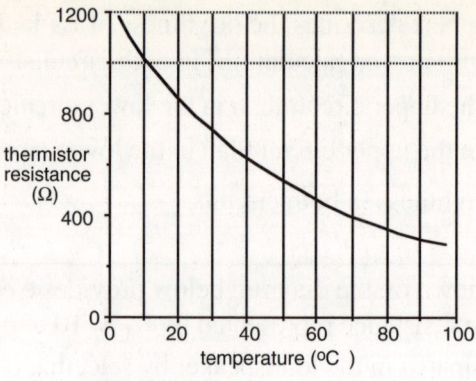

9 Electronics basics

Question 435 *(1998 CAT)*
What is the temperature of the cup of coffee? Show your working.

[3 marks]

The diagram to the right shows a circuit where a light emitting diode (LED) will emit light, or not, depending on the current. The circuit consists of the LED, a variable resistor R and a 12 V battery. The graph shows the current-voltage characteristics of the LED.

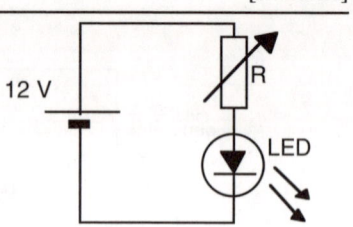

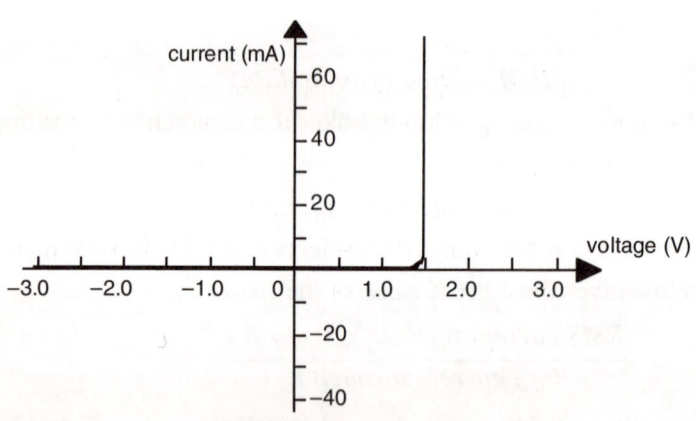

Question 436 *(1999 CAT)*
For R > 150 Ω and R < 750 Ω, the LED emits light and the voltage across it stays the same. Find the voltage across R when its resistance is 350 Ω.

[3 marks]

Question 437 *(1999 CAT)*
Find the current in the LED when the variable resistor is set to 350 Ω.

[3 marks]

The circuits following represent a seatbelt warning system. The switch is the seatbelt. An open switch is an open seatbelt; a closed switch is a closed seatbelt. In circuit C the LED connections are reversed.

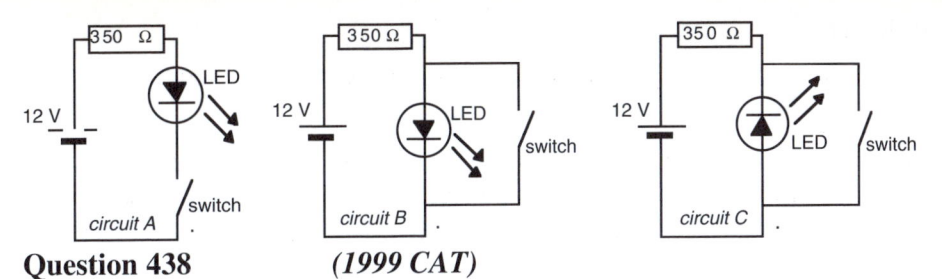

9 Electronics basics

Question 438 (1999 CAT)
Which of the above will ensure that the LED lights only when the seatbelt is open? Explain why your choice is suitable and the other two are not. [4 marks]

Solar cells' convert light energy to electrical energy. Their efficiency depends on the load driven. A typical circuit is shown below.

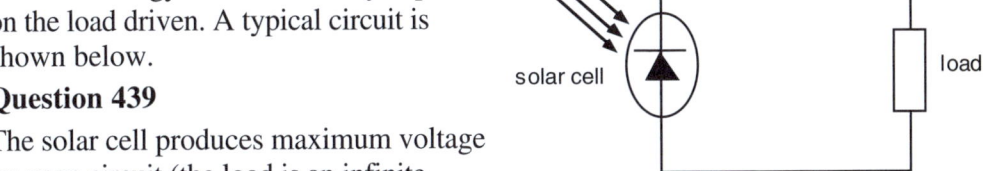

Question 439
The solar cell produces maximum voltage on open circuit (the load is an infinite resistance). What power does the cell deliver on open circuit?

When the load resistance is zero, the cell produces maximum current. The solar cell characteristics are shown following, for one value of light level.

Question 440
What is the maximum current that this solar cell can deliver?

Question 441
When the load is R_1, the current flowing through it is 1.5 mA. What is the voltage across its terminals?

Question 442
What is the power dissipated in the resistor at a current of 1.5 mA?

Question 443
Calculate the value of R_1.

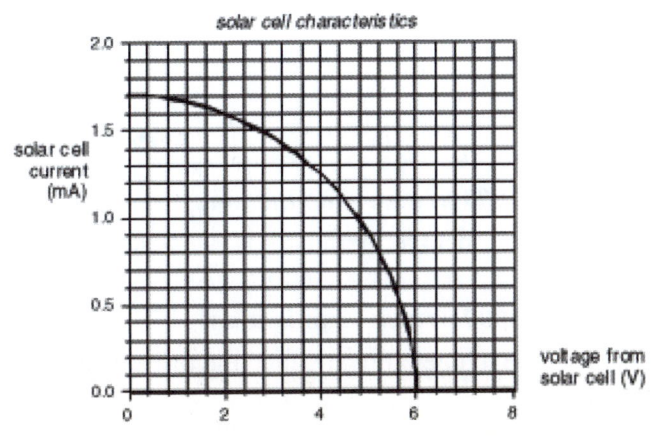

103

9 Electronics basics

Question 444

Mark the approximate point on the characteristics graph above where the power delivered by the solar cell is a *maximum*.

An oscilloscope is used to examine the waveform of the mains voltage from a domestic power outlet. According to the power supply company, the mains voltage is 240 V (RMS), at a frequency of 50.0 Hz. The waveform as seen on the oscilloscope screen is shown on the graph below.

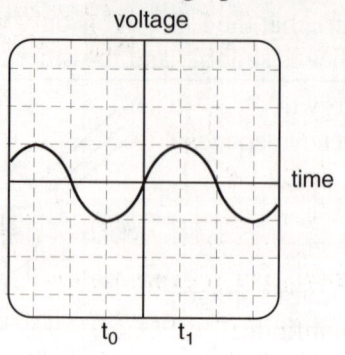

Question 445 (2000 written examination)

What is the time difference (in millisecond) between t_0 and t_1?

Question 446 (2000 written examination)

What voltage does one vertical division of the oscilloscope screen correspond to?

Thermistors are temperature-dependent resistors, and are often used in temperature warning systems. The response function of a thermistor is shown in the graph below.

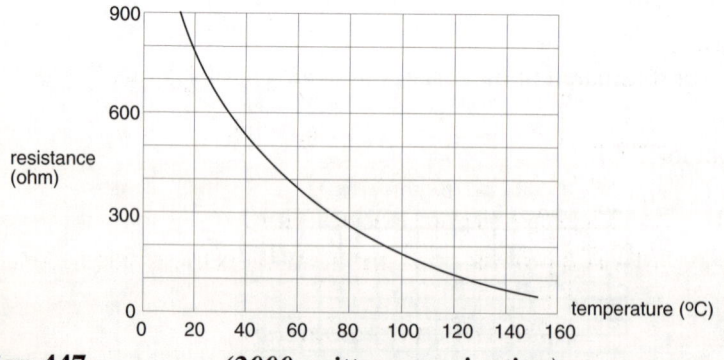

Question 447 (2000 written examination)

Does the graph above suggest that the thermistor is *ohmic* or *non-ohmic*? Explain your answer.

Question 448 (2000 written examination)

From the graph, estimate the temperature at which the resistance of the thermistor will be 400 ohm.

9 Electronics basics

An application for the thermistor is shown below. The buzzer (resistance 100 Ω) produces an audible sound when the current exceeds 10 mA.

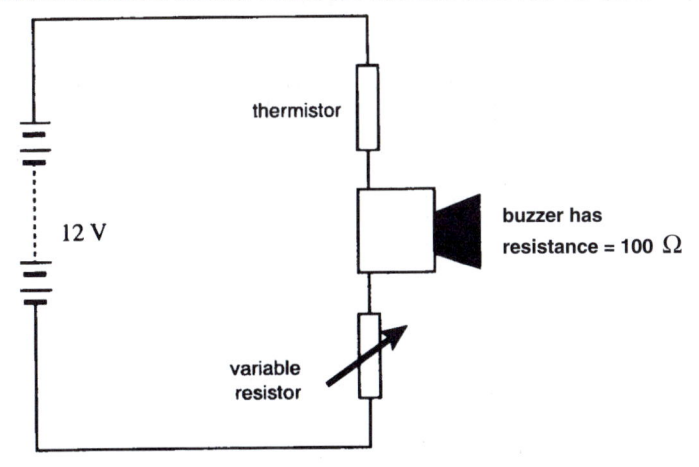

Question 449   *(2000 written examination)*

What is the resistance of the variable resistor required to activate the buzzer, when the thermistor has a resistance of 400 ohms? Show your working.

[3 marks]

A solar cell is a non-linear diode device that is able to provide electrical power when exposed to light. In the test circuit shown below, the power dissipated in the variable load resistor R depends on the light intensity, and the value of R. For this example the graph shown below shows the solar cell characteristic curve current I versus voltage V for the light intensity used.

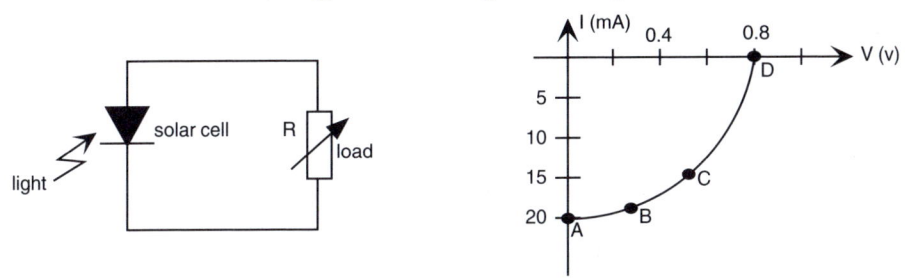

Question 450   *(2001 written examination)*

If R = ∞ ohm, the solar cell provides no current. What is the voltage across resistor R?

Question 451   *(2001 written examination)*

If R is decreased to 0 ohm, the voltage across the solar cell approaches zero. What is the maximum current that the solar cell can deliver to the load resistor R?

Question 452   *(2001 written examination)*

At which point (A, B, C or D) on the characteristic curve above is the largest electrical power dissipated in the resistor R?

CHAPTER 10   Capacitors and diodes

A washing machine has a delay on the door opening after a cycle of washing. Part of this circuit involves the following delay circuit. As the cycle ends switch S closes. At this stage the capacitor is *uncharged*. The door cannot open until it receives a voltage input of at least 8 V.

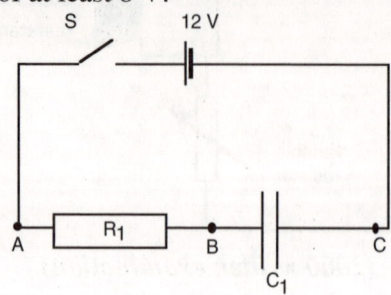

Question 453

Which of the following two points would be the most suitable to connect to for this voltage? Give reasons for your answer.

A   A and B
B   A and C
C   B and C

[3 marks]

Question 454

If R_1 is fixed at 10 kΩ, give a value for C_1 that would give a time delay of around 5 seconds. Show your working.

[3 marks]

Question 455

Sketch the voltage across AB as a function of time, from the time when the switch S closes to when the capacitor is close to fully charged.

[3 marks]

A single diode is used to convert an AC voltage into a half wave rectified voltage using the circuit shown.

Question 456

The voltage across XY has an average DC value of greater than zero, but less than the peak value of the original AC voltage. Explain why.

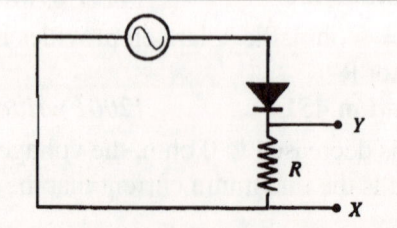

[3 marks]

106

10 Capacitors and diodes

In the circuit shown following the capacitor is fully charged; no current flows in the resistor. The indicator light is lit when the logic output of the AND is 1.

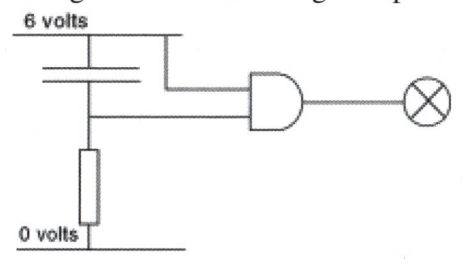

Question 457
Will the indicator light be lit? Give reasons for your answer.

The capacitor has a value of 40 µF, and the resistor has a value of 270 kΩ. A wire is connected for a short time across the terminals of the capacitor and then removed. Whilst the wire is in position all charge is removed from the capacitor.

Question 458
Whilst the wire is in position, is the indicator lamp lit or unlit? Give reasons for your answer.

Question 459
When the wire is removed, the capacitor begins to recharge. Which of the following is closest to time when the capacitor will be ~ 63% charged?

A	1 sec
B	10 sec
C	100 sec
D	10 000 sec

The capacitor in the diagram below is fully charged from a 6V battery. The capacitor has a value of 220 µF and the resistor a value of 110 kΩ.

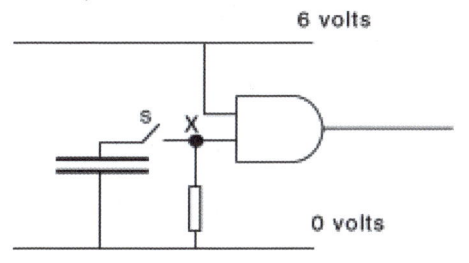

Question 460
Before the switch S is closed, the logic state of the output of the AND gate is 0. Explain why. [3 marks]

Question 461
When the switch S is closed, the voltage at the point X rises to nearly 6 volts for a short time. What is the time constant of the combination of the capacitor and resistor?

Question 462

Explain why the voltage at X rises and also why it drops back to 0 volts after a while. [4 marks]

In the circuit a cathode ray oscilloscope is used to measure the voltage V_C across the capacitor as a function of time. The capacitor is initially uncharged; $V_C = 0$ V. At time t = 0 the switch is set at *A*.

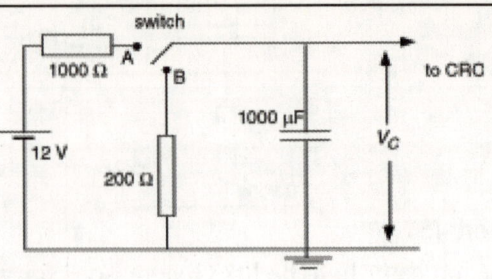

Question 463   *(1996 CAT)*

Which of the graphs below best represents V_C?

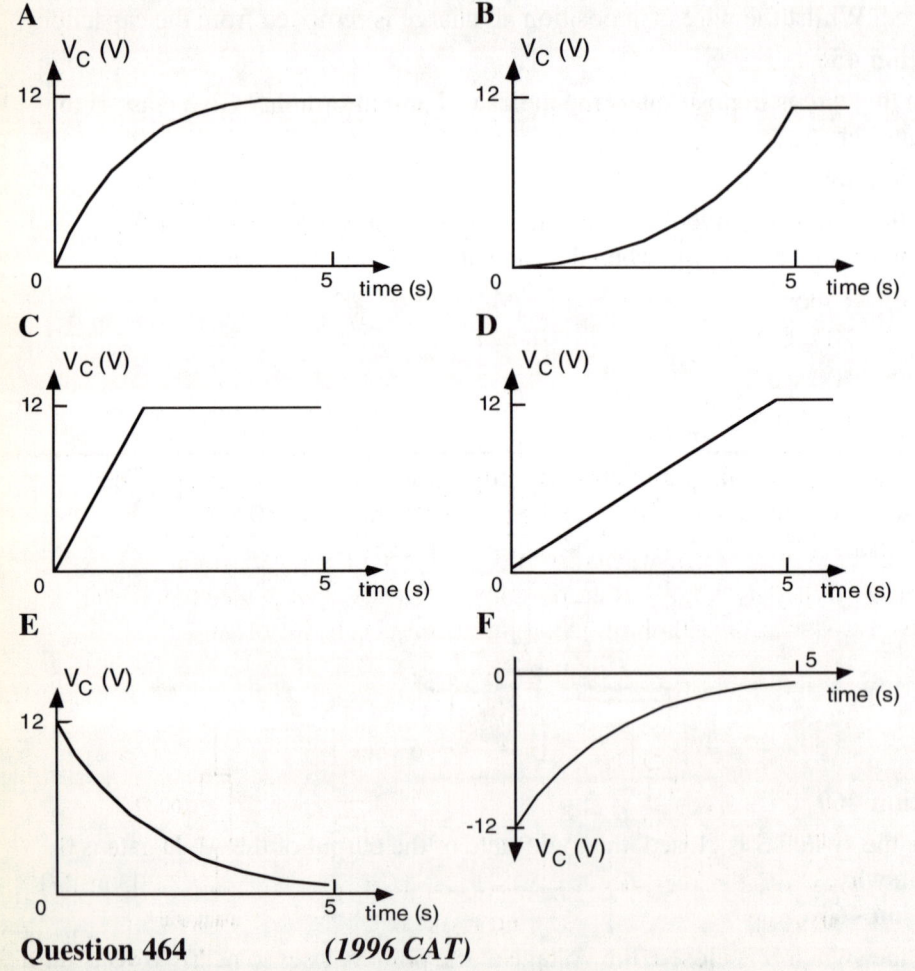

A, **B**, **C**, **D**, **E**, **F**

Question 464   *(1996 CAT)*

When the capacitor is fully charged, with the switch at position *A*, the current in the 1000 Ω resistor is zero. Explain why the current is zero. [2 marks]

108

Question 465   *(1996 CAT)*

At time $t = 100$ s the switch is moved to position B. Which one of the graphs below best represents the voltage across the capacitor?

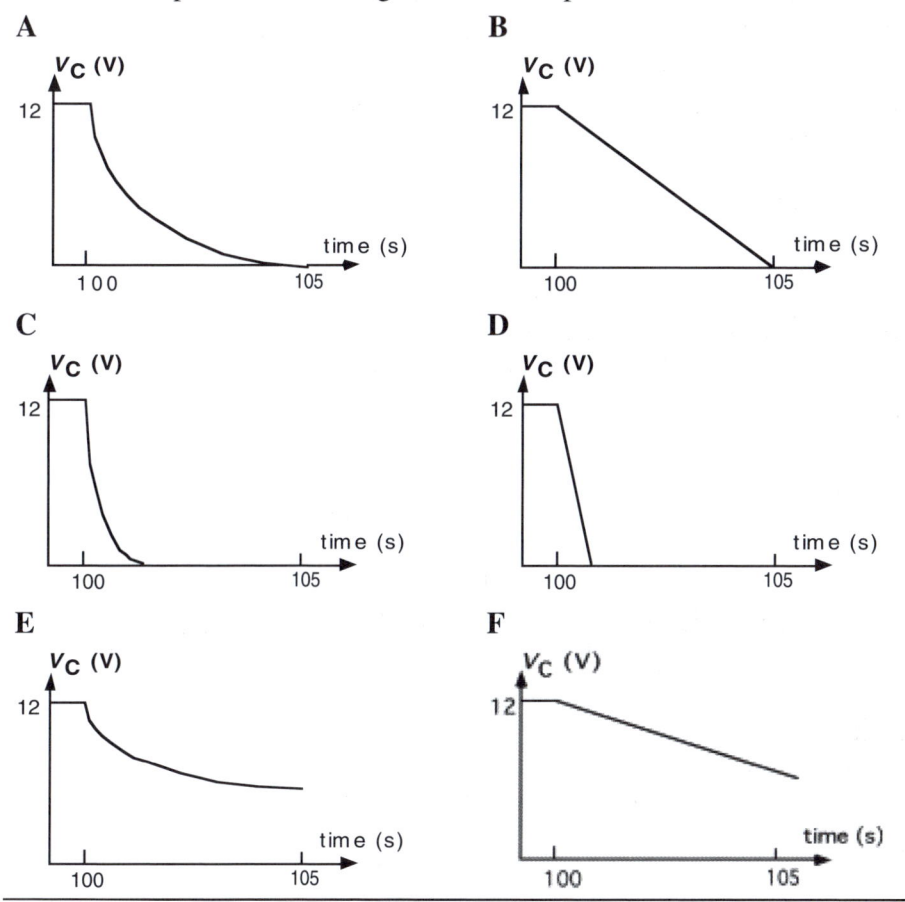

A

V_C (V), 12, time (s), 100, 105

B

V_C (V), 12, time (s), 100, 105

C

V_C (V), 12, time (s), 100, 105

D

V_C (V), 12, time (s), 100, 105

E

V_C (V), 12, time (s), 100, 105

F

V_C (V), 12, time (s), 100, 105

A DC power supply is constructed from a transformer and electronic components. This DC supply is used to power an electrical device that has an effective resistance of 100 ohm. A possible circuit is shown below.

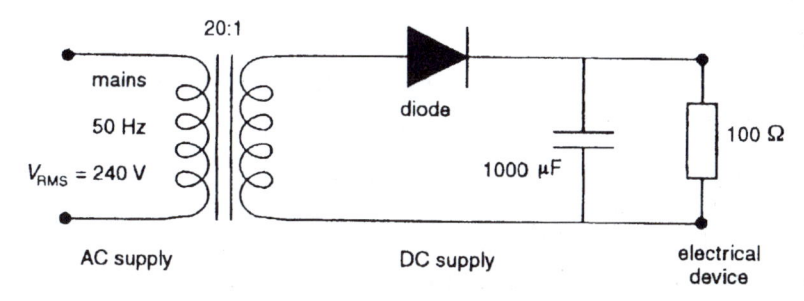

mains 50 Hz, $V_{RMS} = 240$ V, 20:1, diode, 1000 μF, 100 Ω

AC supply — DC supply — electrical device

To produce a DC voltage from an AC supply requires two processes, *rectification* and *smoothing*.

10 Capacitors and diodes

Question 466   *(1997 CAT)*
Explain in words what is meant by *rectification*.

Question 467   *(1997 CAT)*
What is the smoothing time constant of the circuit above? Give your answer in millisecond.

The graph below shows the time variation of the voltage across the secondary windings of the transformer.

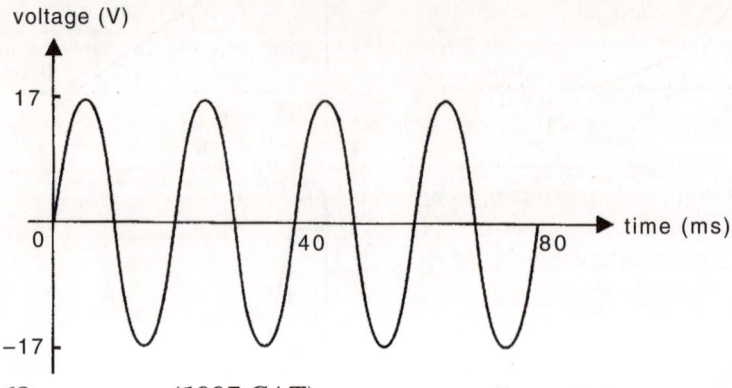

Question 468   *(1997 CAT)*
Sketch a graph of the voltage across the 100 ohm resistor as a function of time. Use the same time scale as the graph above.

The diagram below shows a rectifying/smoothing circuit attached to the output of a voltage generator.

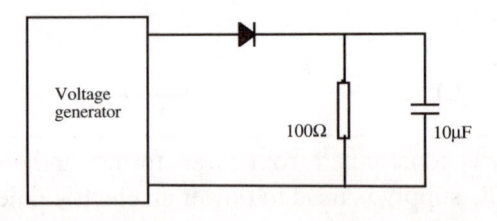

Question 469
What is the time constant of the smoothing section of this circuit?

Question 470
The circuit will give a smooth output (very little ripple) for an output signal of period (one or more answers):

A   10 μs   B   0.1 ms   C   10 ms   D   100 ms

A DC power supply is shown following. It has a half-wave rectifying section (the transformer and the diode) followed by a smoothing section (the capacitor and the load resistor). The load resistor has a resistance of 300 Ω.

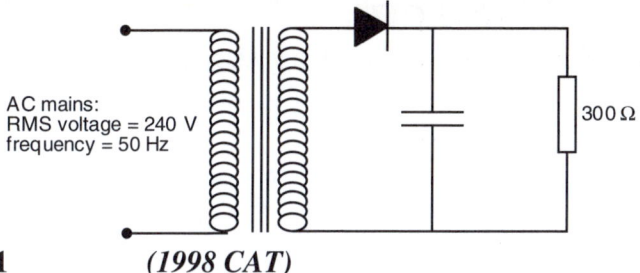

10 Capacitors and diodes

Question 471 *(1998 CAT)*
If the time constant of the smoothing section is 30 ms, what is the capacitance of the capacitor? Show your working and give your answer in microfarad (µF).

Question 472 *(1998 CAT)*
Which of the voltage waveforms below best describes the voltage across the load resistor, V_R? (The dotted line indicates the output voltage of the transformer when the diode and capacitor are removed from the circuit.)

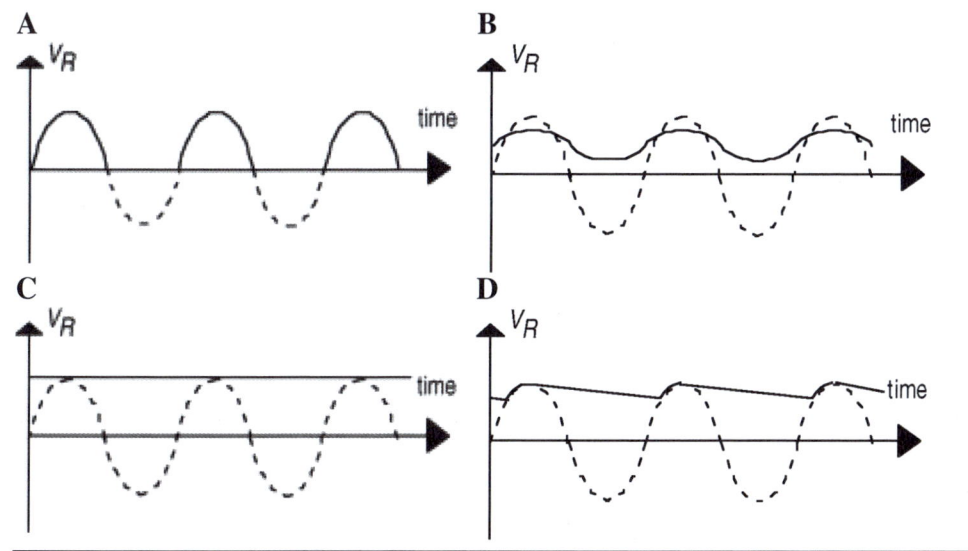

Many electric devices make use of the time taken to charge a capacitor. The diagram shows a simplified charging circuit with a capacitor, a switch, a 1 kΩ resistor and a 10 V battery. When the switch is closed the capacitor starts to charge. The time constant is 1.2 s.

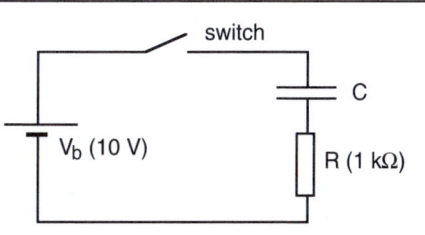

Question 473 *(1999 CAT)*
Calculate the capacitance of this capacitor.

10 Capacitors and diodes

Question 474   (1999 CAT)

Which of the following best shows how the voltage across the resistor varies with time after the switch is closed?

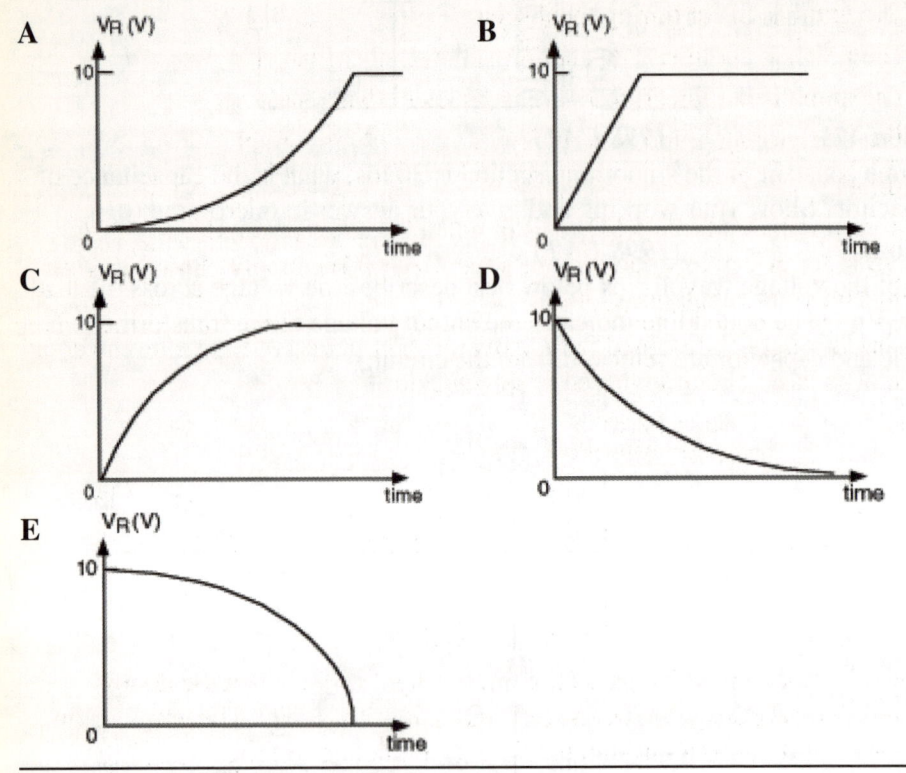

A — V_R (V): curve increasing slowly then rising sharply toward 10
B — V_R (V): rises linearly to 10 then flat
C — V_R (V): rises quickly and levels off at 10
D — V_R (V): starts at 10 and decays exponentially to 0
E — V_R (V): starts at 10 and falls as a quarter-circle to 0

The circuit shown below is being used to measure the charging rate of a capacitor. At the start of the experiment the switch is set to position X. Before this is done the capacitor is *uncharged*.

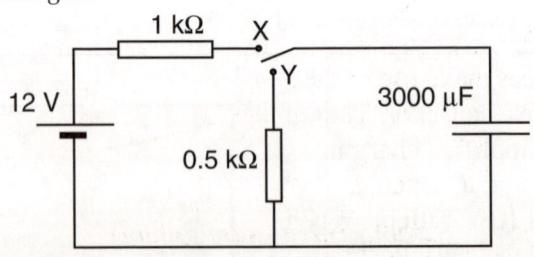

Circuit: 12 V battery, 1 kΩ resistor in series with switch (positions X and Y), 3000 μF capacitor, and 0.5 kΩ resistor.

Question 475

Sketch the voltage across the capacitor as a function of time. The time axis should include *approximate* figures. Take t = 0 as the time that the switch is connected to X.   [3 marks]

Question 476

When the capacitor is fully charged, the current through *both* resistors is zero. Explain why this is so.   [3 marks]

Question 477

Which of the following statements best describes the shape of the discharging curve of the voltage across the capacitor?

A   The graph is linear during the discharge time.
B   The graph is not linear; discharging takes longer than charging.
C   The graph is not linear; discharging is faster than charging.
D   The discharging graph is very close to the charging graph.

Question 478

Would the combination of a 5 kΩ resistor and a 0.1 µF capacitor be a suitable combination for effective smoothing of a rectified 50 Hz supply? Show all your reasoning and your calculations.

A car's interior light is designed to fade over a short period of time after the door is closed. The basic circuit involved is shown below.

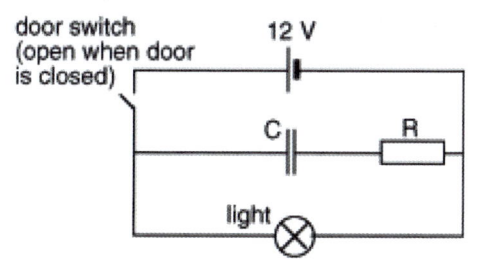

With the door open (*switch closed*) the interior light is *on*. When the door is closed, the *switch opens*, and the interior light *fades gradually*. The graph below shows the current flowing through the light as a function of time.

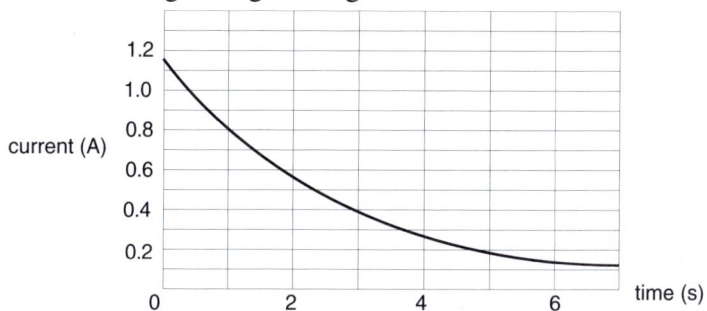

Question 479   (2000 written examination)

Estimate the time constant for this circuit. Give your answer to the nearest 0.2 s.

Question 480   (2000 written examination)

The value of the capacitance, C, in the circuit is 100 µF. What is the value of the resistance R? (You may assume that the resistance of the light can be ignored.)

Modern electronic devices are designed to operate on low DC voltages. The diagram following shows part of a circuit which provides this. It consists of a low-voltage AC input connected to a rectifier.

10 Capacitors and diodes

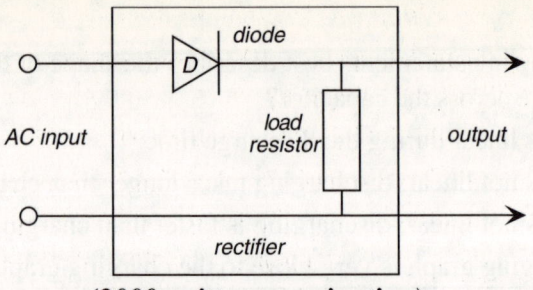

Question 481   *(2000 written examination)*

Which one of the following diagrams best represents the voltage as a function of time, at the output of the rectifier?

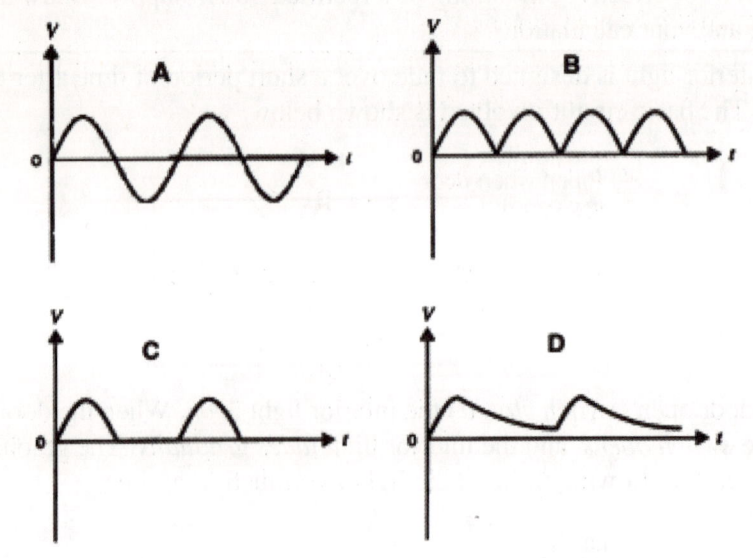

Question 482   *(2000 written examination)*

A capacitor is added to the rectifying circuit to smooth the DC output. Which *one* of the following circuits has the capacitor correctly located?

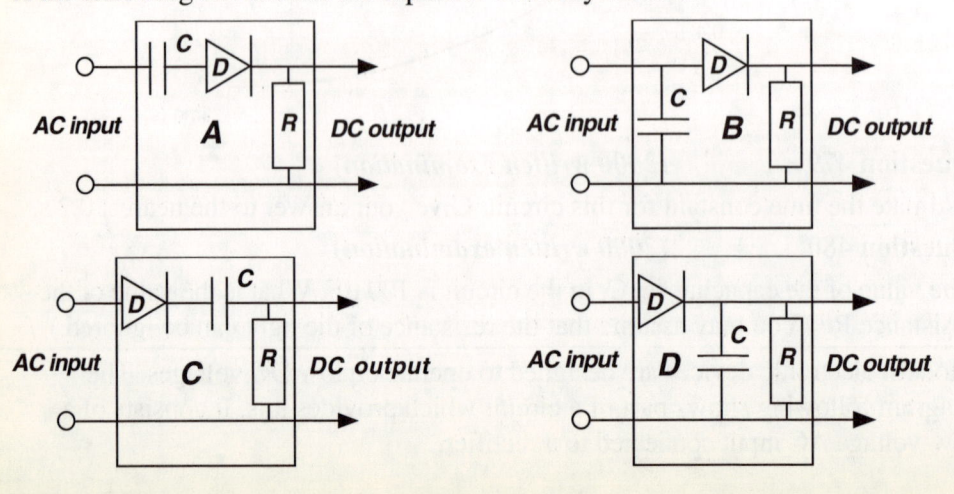

114

Question 483 *(2000 written examination)*
If the resistance of the load resistor R is 100 Ω, which *one* of the following gives the capacitance of a capacitor that will provide the smoothest DC output?
A   1 000 μF   B   1.0 μF   C   1 000 nF   D   100 nF

In modern cars a simple R-C circuit is used to control the length of time that the interior light remains on after the door is closed. This control circuit is shown below. Additional circuitry (which is not shown) arranges for the light to be ON **only** if the voltage across the capacitor (V_C) is greater than 8 V.

When the car door is open, the switch S is connected to the OPEN contact, and the capacitor immediately charges fully, so that V_C = 12 V. When the car door is closed, the switch S moves to the CLOSED position, and the capacitor discharges through the resistor, and the voltage V_C falls to 8.0 V precisely 4.0 s later. The time constant of the RC circuit is 9.87 s.

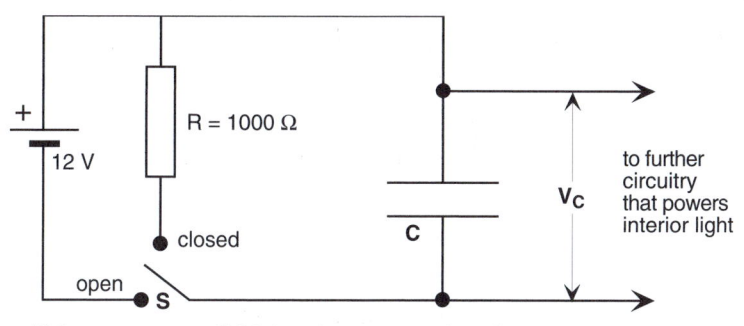

Question 484 *(2001 written examination)*
What is the value of the capacitor C? Include the unit in your answer.

[3 marks]

Question 485 *(2001 written examination)*
If you wanted the light to turn off 6.0 s after the door was closed (instead of 4.0 s), would you increase or decrease the value of the resistor R? Justify your answer.

[3 marks]

Question 486 *(2001 written examination)*
The original circuit is used, with R = 1 000 Ω, but the battery voltage reduced to 10 V. Which one of the statements below gives the best estimate of the time when the light would turn off after the door is closed?
A   after more than 4.0 s
B   at 4.0 s
C   after less than 4.0 s

10 Capacitors and diodes

The left-hand diagram below shows a 20 Hz sinusoidal voltage, V_{in}. It has a **peak-to-peak** voltage of 10 V. The right-hand diagram shows a circuit that is designed to rectify this voltage.

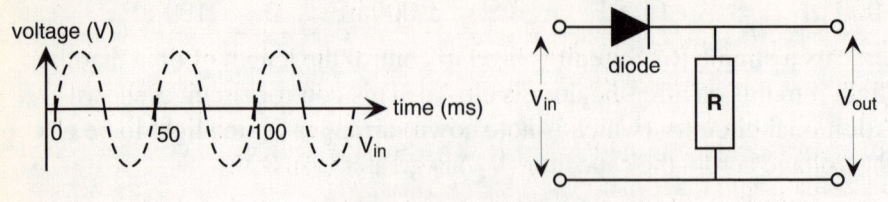

Question 487 *(2001 written examination)*
On the graph above show the correct numbers on the scale of the vertical axis.

Question 488 *(2001 written examination)*
Sketch the form of the output voltage from the circuit shown above. Use the same time and voltage scales on the axes.

To the circuit above is added a smoothing capacitor, in parallel with the resistor R. This modified circuit is shown below.

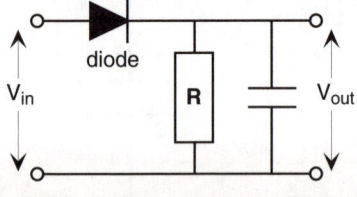

Question 489 *(2001 written examination)*
Sketch the form of the output voltage from this modified circuit, for the same input voltage. You should assume that the smoothing time constant has a value of 100 ms. Once again, use the same voltage and time scales as in the previous questions.

CHAPTER 11 — Amplification

Andrea needs a pre-amplifier for her guitar. She needs it to have a gain of 500, and to have a linear region which can accept signals of up to 4 mV (peak). The graph below shows the shape of the characteristic graph for this pre-amplifier, but it requires numbers on the axes.

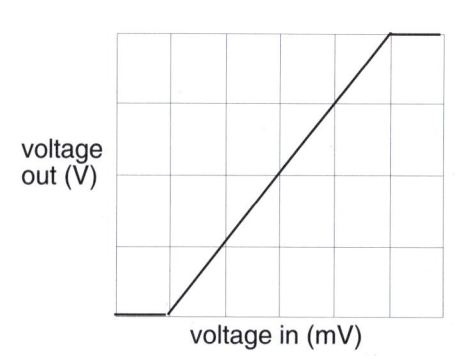

Question 490
Complete the graph by adding the correct numbers to the axes.

To test her pre-amplifier, Andrea feeds the following test signal into it.

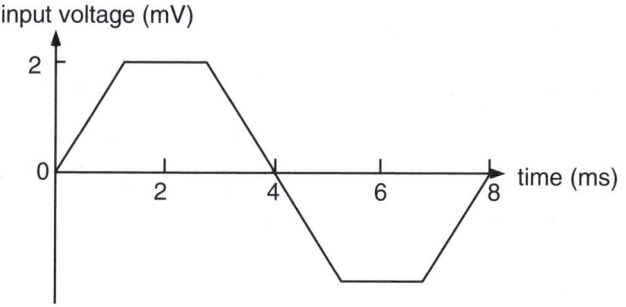

Question 491
Sketch the output signal from pre-amplifier (use the same time scale).

A linear amplifier has input–output characteristics as shown below.

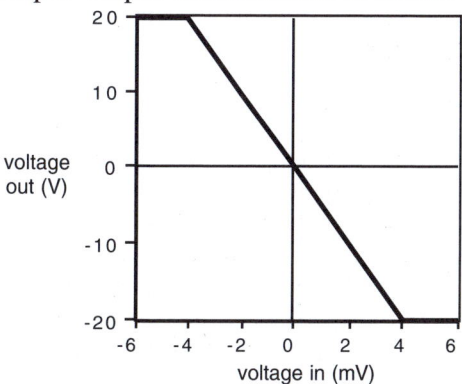

Question 492
Calculate the gain of this amplifier. Indicate in your answer whether it is inverting or non-inverting.

117

11 Amplification

Question 493

The waveform shown below is amplified by the amplifier described above. Sketch the output voltage as a function of time.

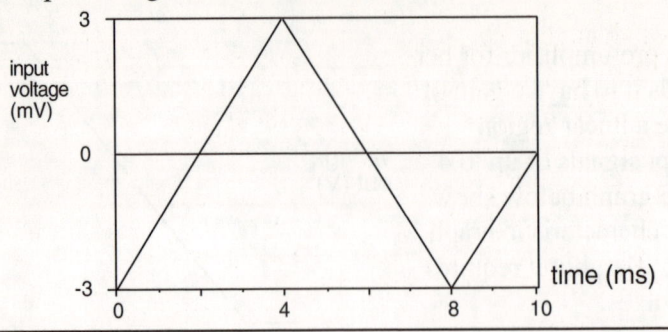

A device has the input–output voltage graph shown. (For input voltages above 1 V the output is 0 V.)

Question 494

The device can operate as a NOT gate. Explain.

Under certain circumstances it can also work as an amplifier.

Question 495

What would be the gain of such an amplifier?

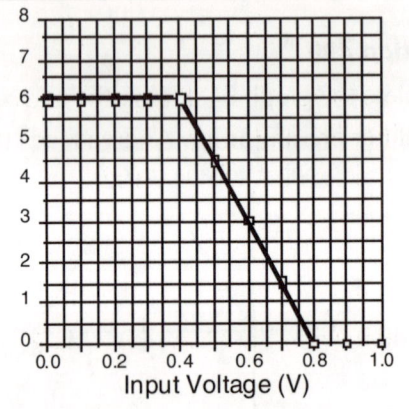

Question 496

What is the largest input AC voltage possible without distortion?

A small amplifier has input–output characteristics as shown following.

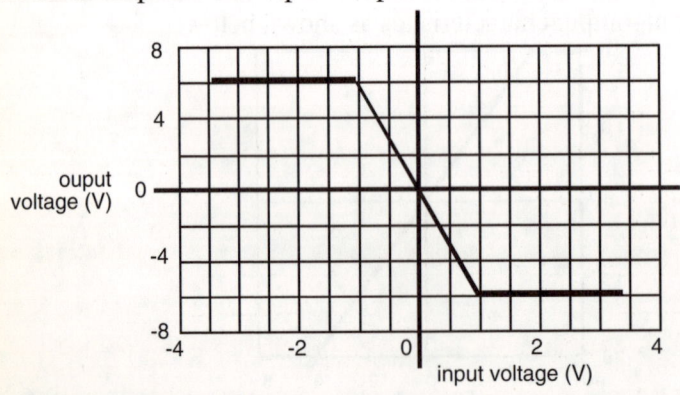

Question 497

What is the gain of the amplifier?

118

Question 498
A sinusoidal AC input signal of 0.8 V$_{peak}$ is used. Sketch this signal.

Question 499
Sketch the output signal graph. [4 marks]

The table shows DC output and input voltages for an amplifier.

V$_{in}$ (mV)	V$_{out}$ (V)	V$_{in}$ (mV)	V$_{out}$ (V)
−60	8	−40	8
−20	4	0	0
20	−4	40	−8
60	−8		

Question 500
Sketch a graph of these values.

Question 501
What is the gain of the amplifier?

Question 502
What is the maximum AC voltage input possible without distortion? Express your answer as a *peak-to-peak* voltage.

The graph below shows a sinusoidal input voltage for the amplifier.

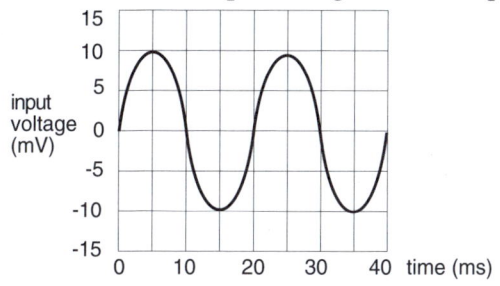

Question 503
Sketch the output graph for this input. Use the same time scale.

In some applications, amplification is carried out by *two* amplifiers, arranged as shown in the diagram below.

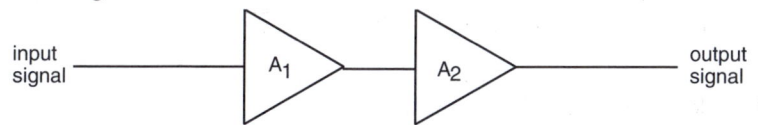

Question 504
In one application of this technique, an input signal of 2.4 mV AC (p-p) is to be amplified to a final output of 3.6 V (p-p). Which of the following values of gain for the amplifiers A$_1$ and A$_2$ will accomplish this?

11 Amplification

	Gain of A_1	Gain of A_2
A	× 3	× 5
B	× 50	× 30
C	× 1.5	× 1
D	× 12	× 100

Question 505

Amplifier A_2 is then replaced by an amplifier that distorts when the input voltage is greater than 48 mV. What is the maximum allowable gain for A_1 in these circumstances?

A key on a new keyboard generates a sinusoidal voltage depending on the average speed of the key whilst being struck. The data below shows the dependence of this voltage on the average speed of a particular key. This voltage is amplified by a voltage amplifier, with a gain of ×150.

average speed of key (m s^{-1})	RMS voltage generated (mV)
0.4	4
0.8	8
1.2	12
1.6	16
2.0	20
2.4	24

Question 506

At what speed must the key move to produce an RMS output of 2.1 V?

Question 507

What is the peak-to-peak value of this voltage?

A linear inverting amplifier has a gain in its linear region of -120. Its linear region is from -12 mV through to $+12$ mV (DC). Outside this region it is non-linear.

Question 508

What is the maximum RMS AC voltage that can be amplified by this amplifier without distortion?

Question 509

Sketch the graph of DC input voltage against DC output voltage for this amplifier, between $V_{in} = -10$ mV and $V_{in} = +10$ mV. Label the axes carefully.

[3 marks]

The graph following shows the relationship between the output voltage and the input voltage for an amplifier.

11 Amplification

Question 510   *(1996 CAT)*

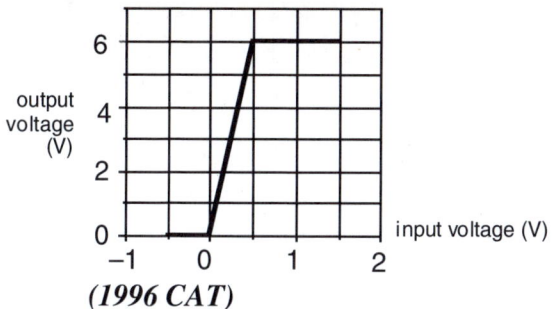

Between $0 < V_{in} < 0.5$ V, what is the gain of the amplifier?

The graph below shows part of an input voltage for this amplifier.

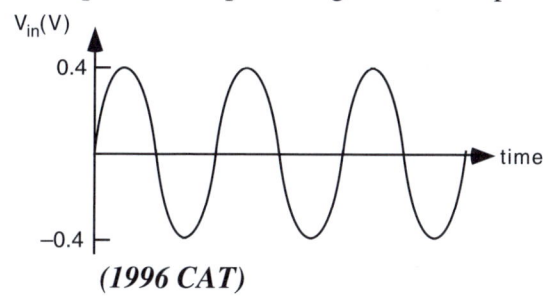

Question 511   *(1996 CAT)*

Which of the following best represents the output of this amplifier if the input voltage is as shown above? All the graphs have the same time scale.

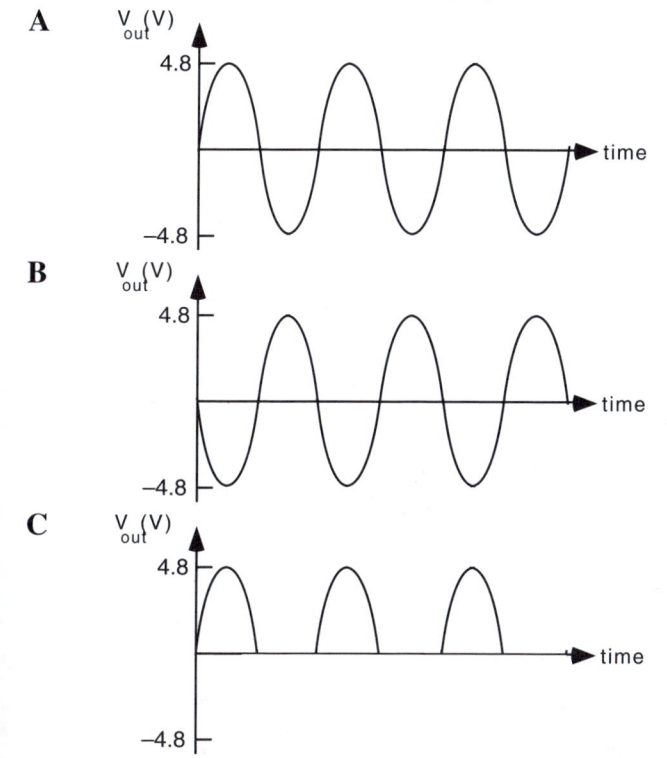

A

B

C

11 Amplification

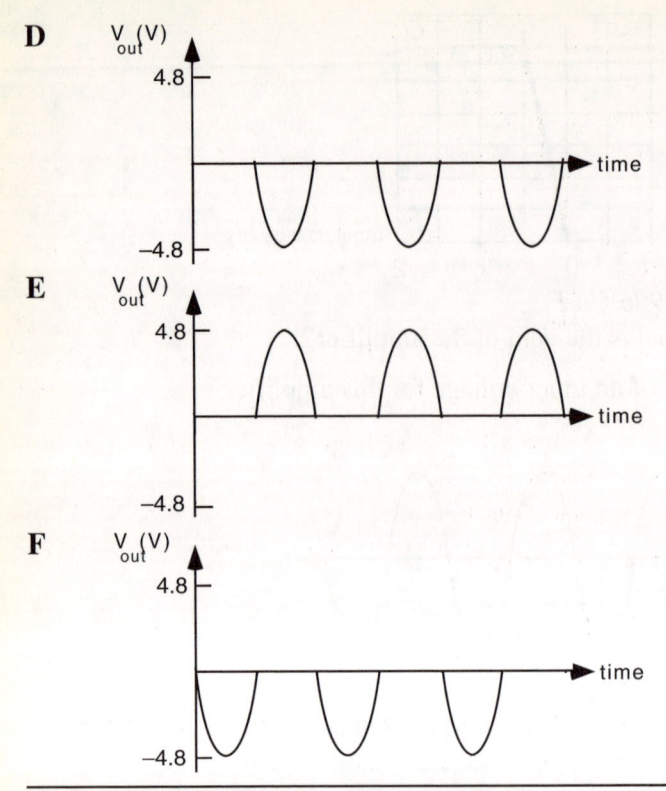

D V_{out}(V), with values 4.8 and −4.8 marked; waveform shows downward-pointing humps (inverted bumps) vs time.

E V_{out}(V), with values 4.8 and −4.8 marked; waveform shows upward-pointing humps vs time.

F V_{out}(V), with values 4.8 and −4.8 marked; waveform shows downward humps reaching near −4.8 vs time.

A student uses an oscilloscope to display both the input voltage V_{in} and the output voltage V_{out} of an amplifier. The input voltage can be of various shapes, with a magnitude varying from 0 to ±2 V peak. When $V_{in} = 0$ V, $V_{out} = 0$ V. The oscilloscope display for both input and output voltages is shown for two experiments (1 and 2) in the following diagrams. In each experiment the vertical scale is 2 V/division and the position of zero volts is indicated for both input and output traces by dashed lines.

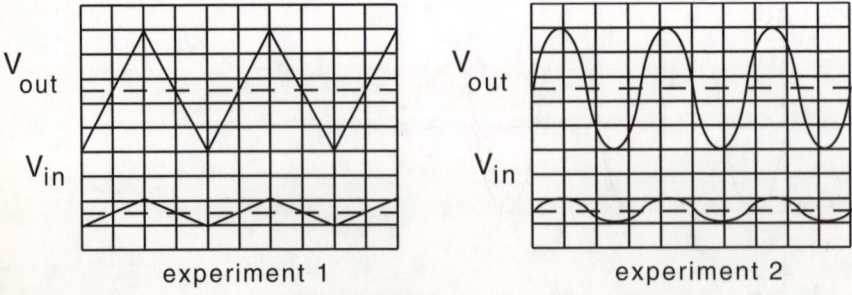

experiment 1 — V_{out} triangular wave, V_{in} triangular wave

experiment 2 — V_{out} sinusoidal wave, V_{in} sinusoidal wave

Question 512   *(1997 CAT)*

Sketch the graph of V_{out} as a function of V_{in} for this amplifier, between $V_{in} = -2$ V and $V_{in} = +2$ V. Label the V_{out} axis with the correct numbers.

Question 513   *(1997 CAT)*

Calculate the gain of the amplifier.

A voltage amplifier produces linear amplification of -200 of AC signals of magnitude up to 30 mV (RMS).

Question 514
Find the *peak-to-peak* output when the input is 23 mV (RMS).

Question 515
Sketch the graph of V_{out} as a function of V_{in} for this amplifier, between $V_{out} = -6$ V and $V_{out} = +6$ V. Label both axes with the correct numbers. Assume it is *non-inverting*.

A voltage amplifier is connected between an AC signal generator and a cathode ray oscilloscope. The arrangement is shown below. The signal generator output is sinusoidal with a peak-to-peak voltage of 50.0 mV.

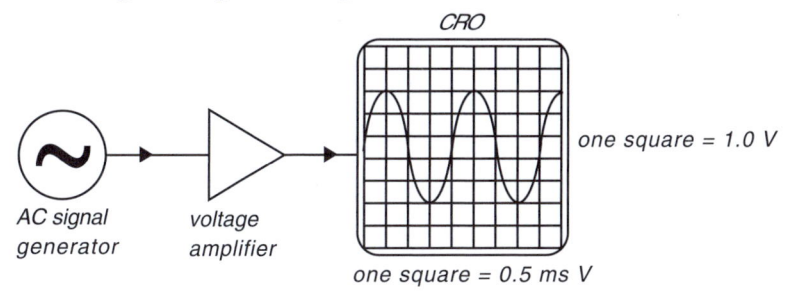

Question 516 (1998 CAT)
Find the RMS output voltage of the signal generator. Show your working.

Question 517 (1998 CAT)
Find the gain of the amplifier. Show your working.

Question 518 (1998 CAT)
Calculate the frequency of the output voltage of the amplifier.

The input–output voltage characteristics of an amplifier are shown below.

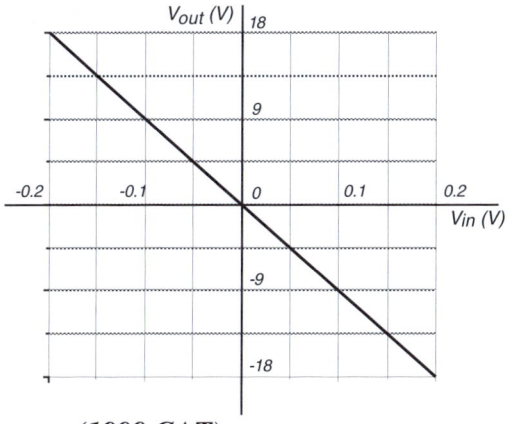

Question 519 (1999 CAT)
What is the voltage gain of this amplifier?

11 Amplification

The graph following shows part of an input voltage to the amplifier.

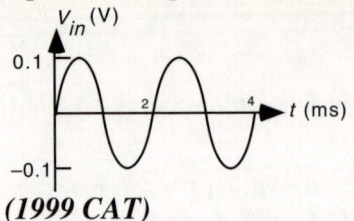

Question 520 (1999 CAT)
Which graph below best shows the resulting output voltage?

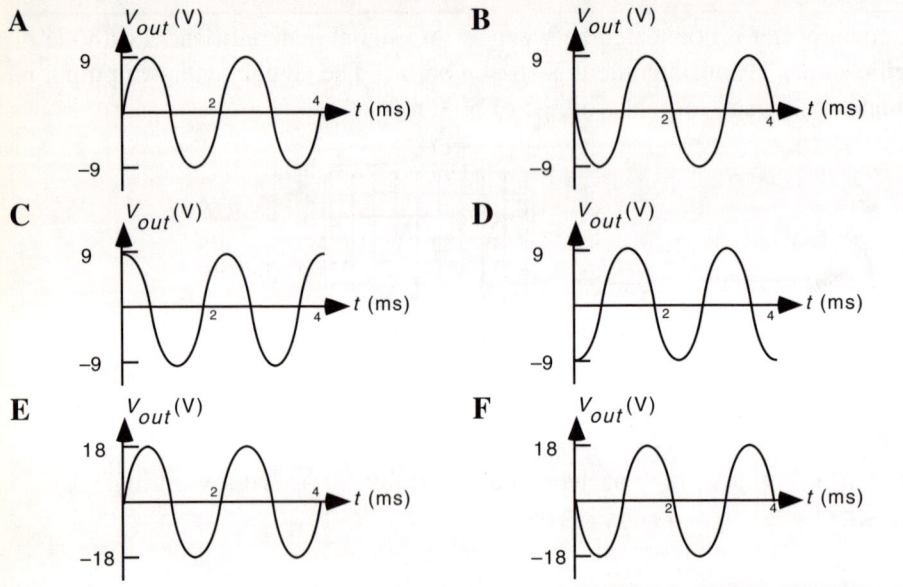

A, B, C, D, E, F

The graphs to the right show the input and the output of a voltage amplifier.

Question 521
What is the gain of the amplifier?

Question 522
What is the signal frequency?

Question 523
What is the RMS output voltage?

Question 524
The current output of the amplifier is 10 mA (peak). What is the RMS output power?

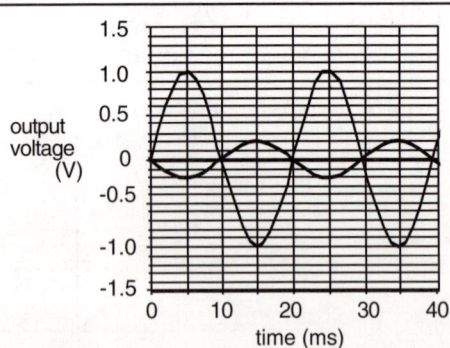

11 Amplification

The graph below shows the input–output characteristics of a DC amplifier.

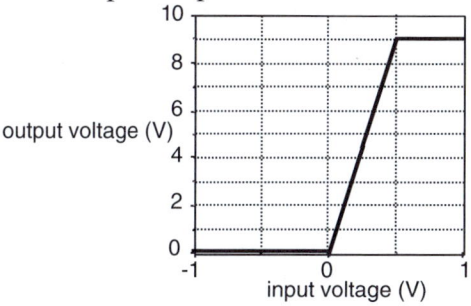

Question 525
What is the gain of this amplifier in the linear region? State whether the amplifier is inverting or non-inverting.

A burglar alarm consists of an oscillator producing a sinusoidal waveform, followed by an amplifier and a speaker. The peak-to-peak voltage of the oscillator is 2.5 V at a frequency of 1200 Hz.

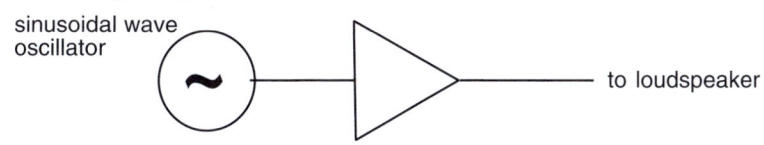

Question 526
What is the RMS value of the voltage from the oscillator?

Question 527
What is the period of the signal from the oscillator?

Question 528
If the loudspeaker is damaged by peak-to-peak signals over 14 V, which of the following gains would be safe to use? (One or more answers.)

A × 6   B × 5   C × 4   D × 3

The first section of a portable tape player amplifies the signal from the tape head. The graph below shows the input and output signals of the amplifier as displayed on an oscilloscope.

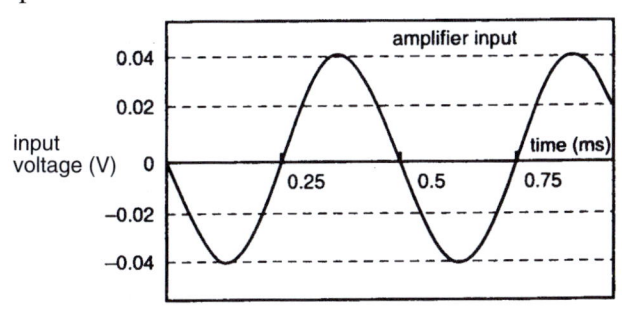

11 Amplification

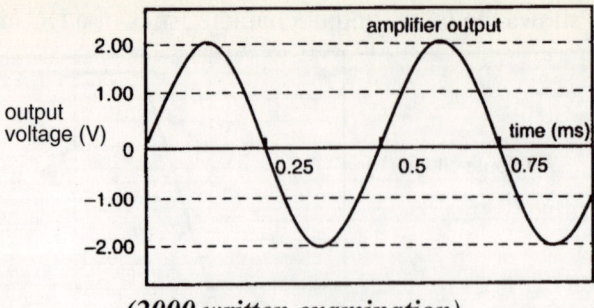

Question 529 *(2000 written examination)*

What is the magnitude of the gain of the amplifier?

Question 530 *(2000 written examination)*

Which one of the diagrams below best indicates the input/output characteristics of the amplifier?

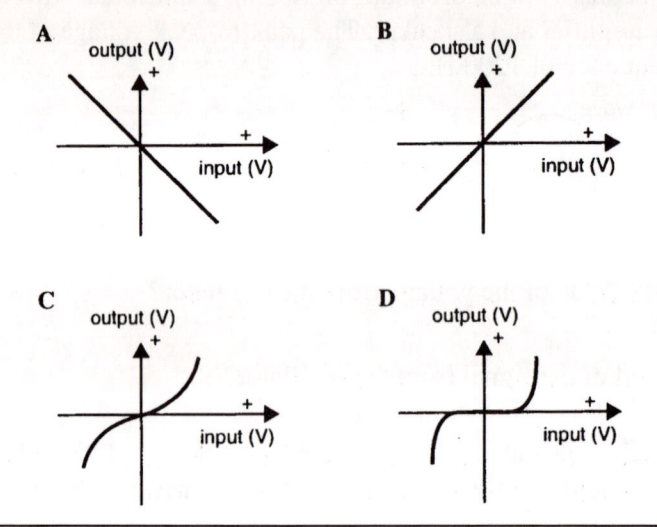

CHAPTER 12 Logic

Question 531
Complete a truth table for the following combination of gates.

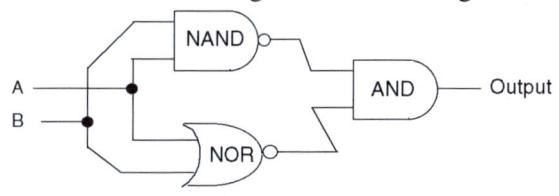

In the logic circuit below, the globe lights when on a logic signal 1.

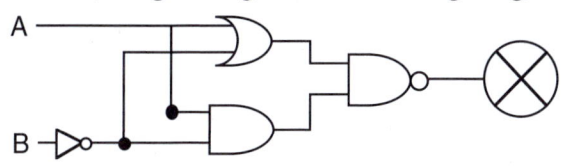

Question 532
Which values of A and B will turn the light on?

A safety door in a production plant is operated by two buttons. They must be pressed at the same time. The door will not open if there is someone on a pressure switch outside the door. The two buttons and the pressure switch all give a HIGH when pressed. The door mechanism needs a LOW to open it. The plant engineers propose the logic circuit shown below.

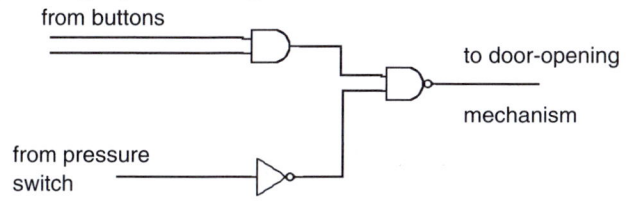

Question 533
Is it suitable for the task? Justify your answer.

[3 marks]

A television quiz show requires contestants to press a button to ring a bell. The first to ring the bell gets the first chance to answer a question. The circuit below is proposed to assist in deciding who pressed their button first. Two NAND gates are involved.

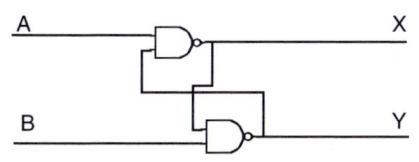

127

12 Logic

Question 534
Complete the spaces left in the truth table following.

Input A	Input B	Output X	Output Y
0	0	1	1
0	1		
1	0		

The bells used operate on a logic state of 0, and the buttons give a logic state of 1 for a very short time, whether they are pressed for a short time or a long time.

Question 535
Which of the following is the correct way to connect the bells?

A   Connect the bell for contestant A to output X directly.
B   Connect the bell for contestant B to output X directly.
C   Connect the bell for contestant A to output X via a NOT gate.
D   Connect the bell for contestant B to output X via a NOT gate.

Question 536
When both contestants press their buttons at very nearly the same time, the first one to press will have an effect on the logic state of the system and the second one will have no effect. Explain why this is so in terms of the logic of the NAND gates. [4 marks]

At a fundamental level in electronic calculators, numbers like 0 and 1 need to be added. The logic can be described by the truth table below.

Input A	Input B	Output X	Output Y
0	0	0	0
0	1	0	1
1	0	0	1
1	1	1	0

Thus $0 + 0 = 0$; $0 + 1 = 1$; and $1 + 1 = 10$ (a digital shorthand for 2). The circuit below is suggested as a way of carrying out this task.

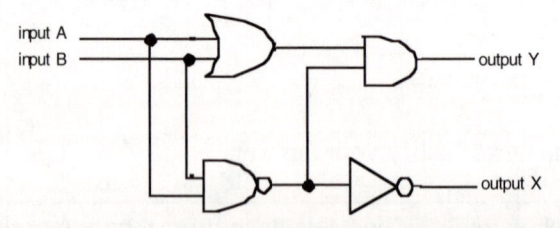

Question 537
Complete the spaces in the table below to verify the logic of the circuit.

A	B	Output of OR	Output of NAND	Output X	Output Y
0	0				
0	1				
1	0				
1	1				

[4 marks]

The combination of gates shown below is sometimes useful.

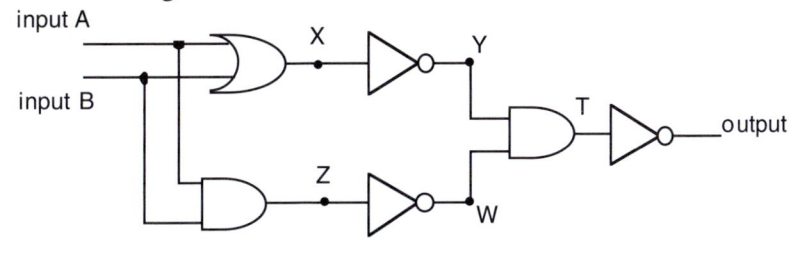

Question 538
Complete its truth table.

Question 539
The circuit shown above can be simplified by removing some of the NOT gates and replacing some of the two input gates with others. Suggest one way this could be done. You may use any of the two input gates NAND, AND, NOR and OR in your replacements.

It is possible to build any of the gates AND, NAND and OR using only NOR gates.

Question 540
Complete the truth table below to decide which of the three gates AND, NAND and OR the following circuit corresponds to. Note that the right hand NOR has one of its inputs permanently connected to 0 logic state.

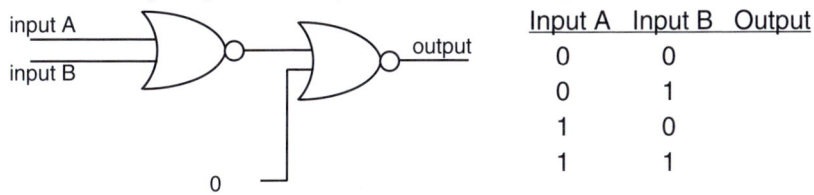

Input A	Input B	Output
0	0	
0	1	
1	0	
1	1	

Question 541
Which of the three gates AND, NAND and OR does the circuit below correspond to? Note that one input of two of the NOR gates is permanently connected to 0 logic state.

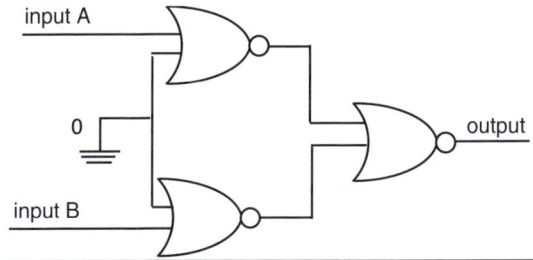

129

Question 542

Which two of the gate combinations in the table following perform exactly the same logical task?

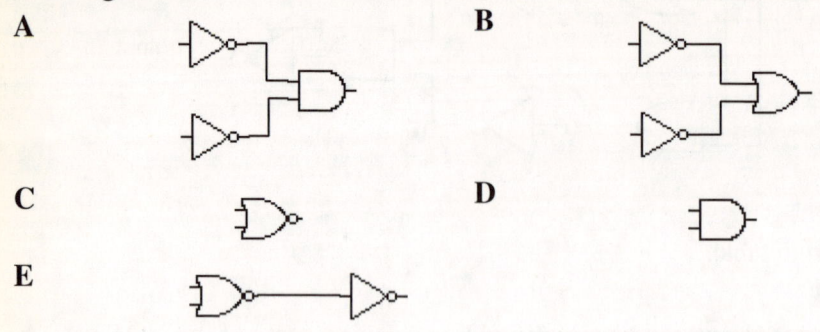

A: (two NOT gates feeding into an AND gate, output inverted — NAND of two NOTs)

B: (two NOT gates feeding into an OR gate, output inverted)

C: (NOR gate)

D: (AND gate)

E: (OR gate followed by NOT gate)

In a food processing plant the mass of food cans is measured electronically as they travel along a conveyor belt. As each can approaches the weighing point it breaks a light beam. The *light sensor* provides a voltage $V_L = 5$ V when a can breaks a beam, making the detected light intensity zero. When the beam is not broken by a can $V_L = 0$ V. The *mass sensor* produces a voltage $V_M = 5$ V when the mass of the can is correct and 0 V when it is too low. The arrangement is shown in the diagram following. The graphs show the voltages as a function of time as two cans move past the sensors.

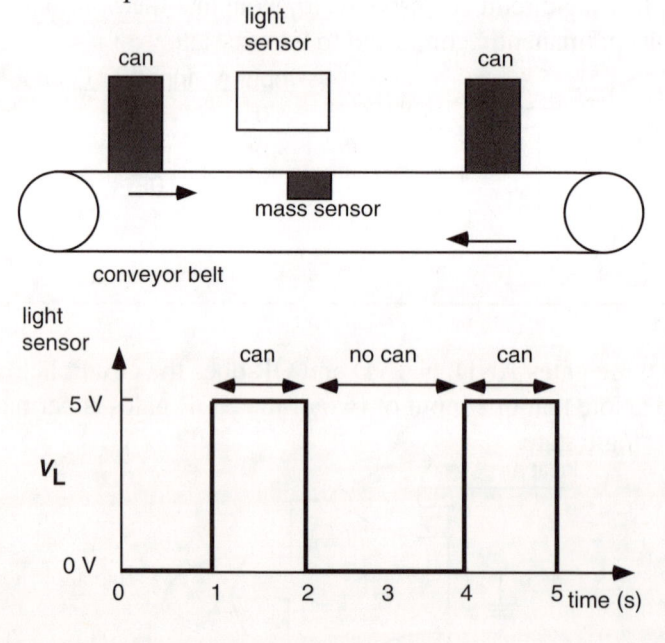

12 Logic

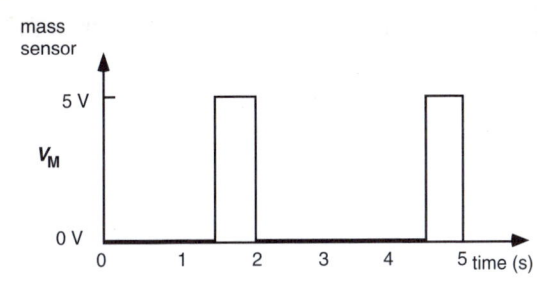

The voltages V_L and V_M are used as the two inputs of an AND gate. The logic states of the two inputs and the output of the AND gate are: 0 V is logic level 1; 5 V is logic level 1.

Question 543 *(1996 CAT)*

Sketch the output voltage of the AND gate as a function of time.

Question 544 *(1996 CAT)*

The output of this AND gate is connected to an indicator light on a control panel. When the output of the AND gate is 5 V the indicator light is *on*. What can be said about the cans if the indicator light is on?

A logic circuit is required to detect cans with less than the correct mass. This logic circuit will need to have:

V_{out} logic level of 1 when a can is present with *less* than the correct mass
V_{out} logic level of 0 when a can is present with the *correct* mass, and
V_{out} logic level of 0 when a can is *not present*.

Question 545 *(1996 CAT)*

Write out the truth table for the required logic circuit.

Question 546 *(1996 CAT)*

Which one of the logic circuits below will provide the required logic?

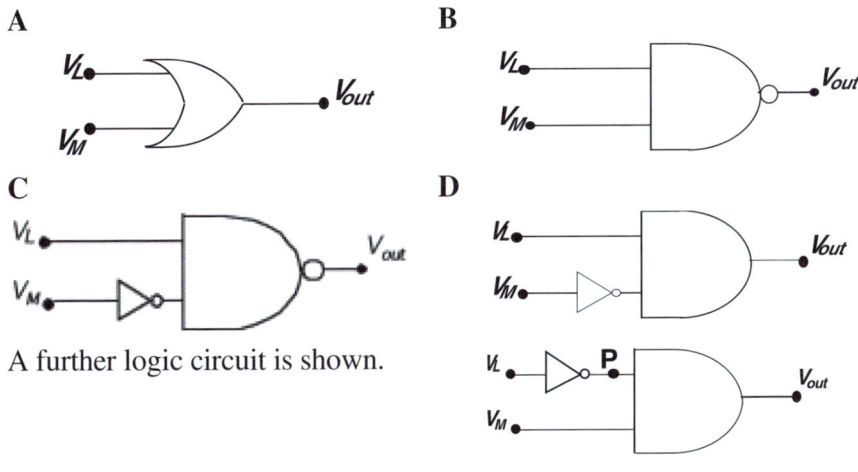

A   B

C   D

A further logic circuit is shown.

Question 547 *(1996 CAT)*

What is the voltage and logic state of the point P when $V_L = 0$ V?

131

12 Logic

Two lights in a control room are used to monitor the status of a moving lift. The two lights are labelled *direction and door*. The conditions that control these lights and their related logic states are defined below.

status light	condition	light	logic state
direction	lift moving up	OFF	0
	lift moving down	ON	1
door	door fully closed	OFF	0
	door open	ON	1

Question 548 (1997 CAT)

Draw a truth table for a circuit that provides a warning in the control room, using an output logic state of **1** if the lift is moving in *either direction* with the door *open*. For all other conditions the circuit should provide an output logic state of **0**.

Question 549 (1997 CAT)

Which logic circuit below provides an output logic state of **1** if the lift is moving down normally with the door closed and an output logic state of **0** otherwise?

An additional status line called LOAD is logic status **1** if the lift is overloaded and **0** if the load in the lift is acceptable. The following question refers only to the LOAD and DOOR logic states.

Question 550 (1997 CAT)

The logic circuit shown indicates a problem with the lift when its OUTPUT is logic state **1**. State, in words, the cause of the problem.

To illustrate the idea of logic gates, a student constructs four circuits, each comprising a DC battery, a resistor, two switches and a light. For all four circuits, the logic states for the switches (S) and the light (L) are defined below.

CONDITION	STATE
SWITCH OPEN	0
SWITCH CLOSED	1
LIGHT OFF	0
LIGHT ON	1

Question 551 (1998 CAT)

Which of the following best corresponds to the circuit where the state of the light (as controlled by the two switches) is correctly described by AND gate logic?

12 Logic

A

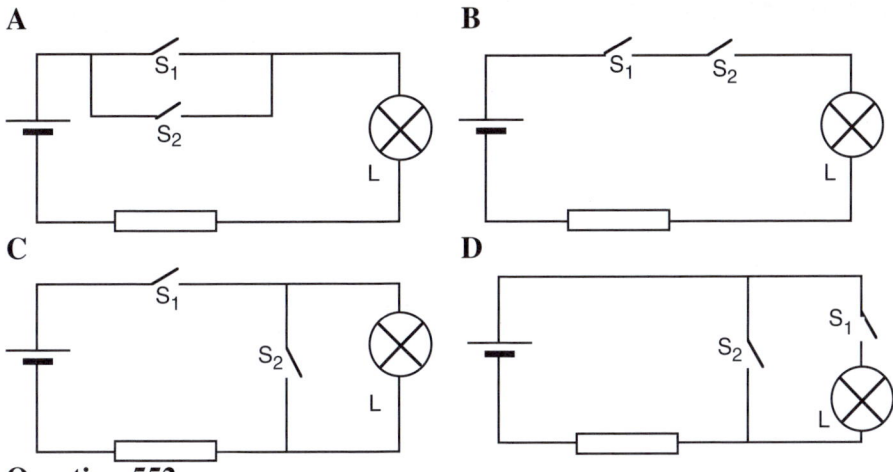

B

C

D

Question 552
Which of the above arrangements corresponds to an OR gate?

Question 553
Write out the truth tables for the gates which are *not* the answers to Questions 551 or 552. [4 marks]

In large rooms the electric light can be operated with two switches, S_1 and S_2. This allows the light (L) to be changed from OFF (logic 0) to ON (logic 1) or from ON to OFF, by changing the position of either of the switches. The switch positions are UP (logic 0) or DOWN (logic 1). Initially both switches are UP; the light is OFF. This logic is shown in the first row of the truth table following.

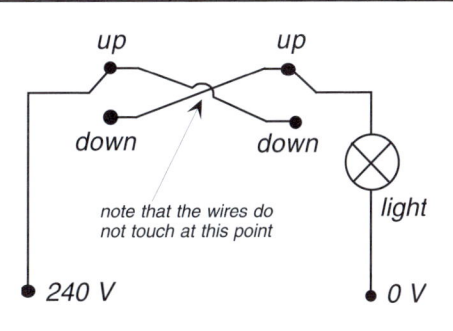

S1	S2	L
0	0	0
0	1	
1	0	
1	1	

Question 554 (1998 CAT)
Complete the right-hand column of the truth table above. [3 marks]

A logic circuit is designed to maintain a constant temperature in a swimming pool. It switches on a heater if the water temperature drops below 27 °C. The automatic heating system operates only when the main switch is ON.

The logic states of the temperature sensor T are defined as:
- logic state 1: temperature is 27 °C or above; logic state 0: temperature is below 27 °C

12 Logic

The logic states of the main switch S are defined as:
- logic state 1: main switch is ON; logic state 0: main switch is OFF

The logic states of the output controlling the heater H are defined as:
- logic state 1: heater is ON; logic state 0: heater is OFF

Question 555 *(1999 CAT)*

Complete the right hand column of the truth table below to indicate the way the two inputs should control the output.

INPUT 1 temp sensor T	INPUT 2 main switch M	OUTPUT heater H
0	0	
0	1	
1	0	
1	1	

[3 marks]

Question 556 *(1999 CAT)*

Which logic circuit below would ensure that the heater turns on when and only when the temperature drops below 27 °C *and* the main switch is ON?

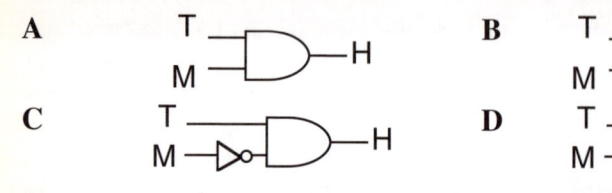

A. T, M → NAND → H

B. T, M → AND → H

C. T, M (M inverted) → NAND → H

D. T, M (M inverted) → AND → H

[3 marks]

The use of electronic circuits to monitor many aspects of our lifestyle is common today. The circuit below is an example of an electronic logic circuit.

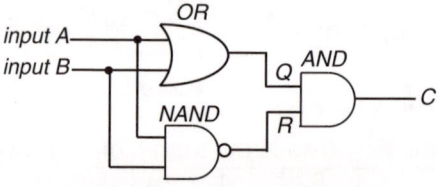

input A, input B → OR → Q; → AND → C; input A, input B → NAND → R

Question 557 *(2000 written examination)*

Complete the columns Q and R of the truth table following for the above circuit.

Input A	Input B	Q	R	Output C
0	0			0
0	1			1
1	0			1
1	1			0

[4 marks]

Two light sensors are used to form an intelligent monitor for a swimming pool gate. As shown on the diagram below, one sensor is placed very close to the ground, and the other at a height of 1.2 m above the ground. Each of the sensors provides a logical output. Logic 0 indicates that the beam has been broken. Logic 1 indicates that the beam is undisturbed.

12 Logic

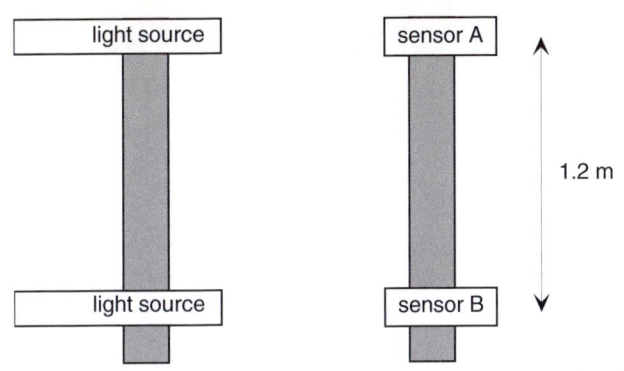

The gate sensors are connected to inputs A and B of the logic circuit shown in the previous question, and output C is connected to an alarm, as shown in the diagram below. The alarm will sound if the output at C is logic 1.

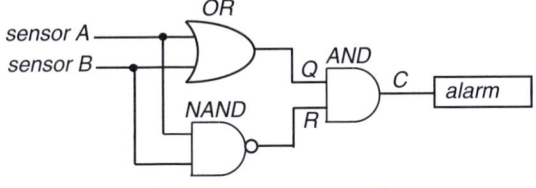

Question 558 *(2000 written examination)*

Which *one or more* of the following situations will sound the alarm.
A   An adult enters the gate.
B   A small child enters the gate.
C   No-one enters the gate.
D   A child jumps the lower beam but breaks the upper beam.

One type of medical scanner works on the principle of the simultaneous detection of 2 γ-rays arriving from a point P in the patient who is placed between two γ-ray detectors. When a γ-ray is detected by either the left (L) or right (R) detector, an output pulse of +5 V and 10 μs duration is provided by the detector. These L and R detector pulses are the inputs to an AND gate. This is shown schematically below.

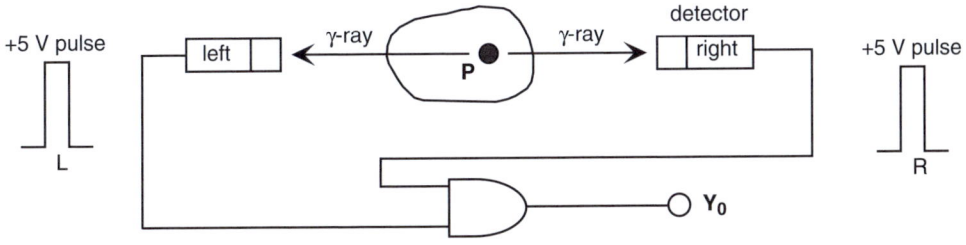

Question 559 *(2001 written examination)*

The diagram below shows a series of signals from the L and R detectors. Complete the third graph for the AND-gate output Y_0 that shows the simultaneous γ-ray events

12 Logic

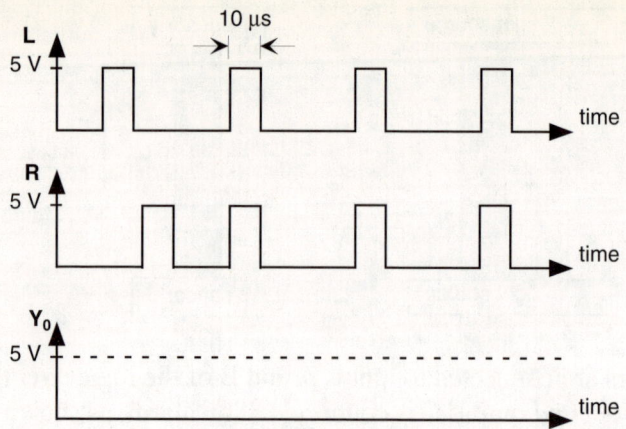

The following diagram shows a more complicated circuit that not only detects simultaneous events via output Y_0, but also detects γ-ray events that **do not** occur simultaneously.

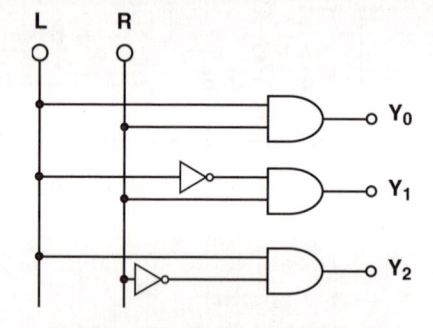

Question 560 *(2001 written examination)*

Complete the timing diagram shown below for the outputs Y_1 and Y_2.

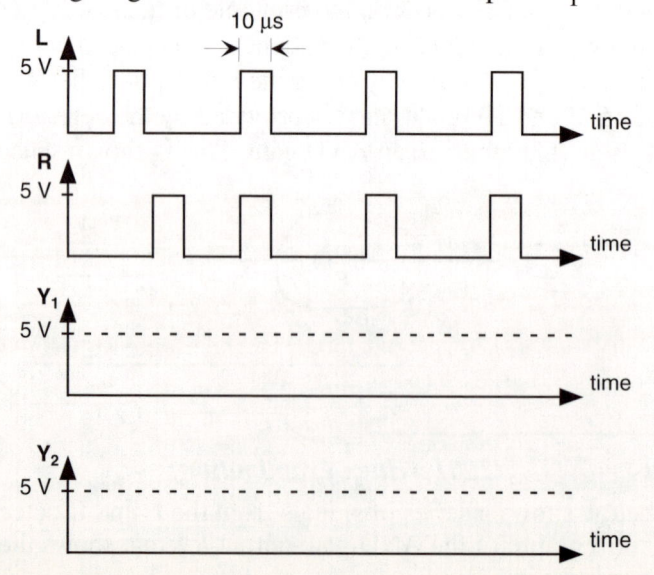

CHAPTER 13 — Flip-flops

Three flip-flops are connected together as in the diagram below.

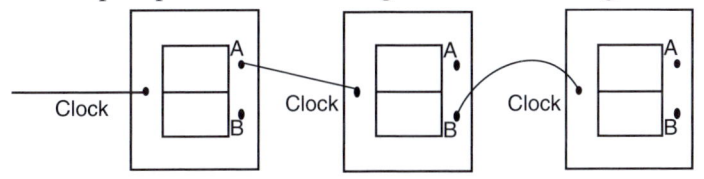

The flip-flops work as follows.
- When A is in logic state 1, B is in logic state 0.
- When A is in logic state 0, B is in logic state 1.
- when a *rising* pulse hits the clock input, A and B swap states.

Question 561
The initial state of the A outputs are 0,0,0. What will be their states if the first clock input is made '1' and '0' 5 times? (after 5 pulses)

Question 562
A 256 Hz square wave is fed into the first input of the flip-flops above. What frequency square wave comes out of input A of the last flip-flop?

Question 563
How many different arrangements of the outputs are possible with three flip-flops arranged as above? *(For example, 0,0,0 and 0,0,1 are two possible arrangements, and there are others.)*

A safety mechanism in a lift counts people entering it. When a person enters the lift a device produces a short voltage pulse as shown below.

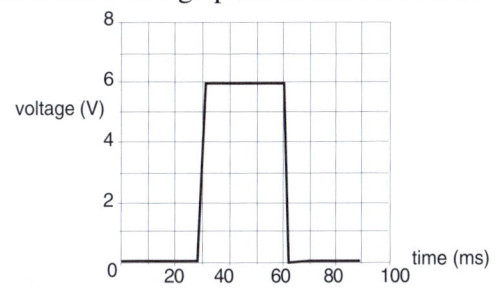

The pulse goes to the first clock input of a series of flip-flops (shown following). The flip-flops change state on a *rising* clock input.

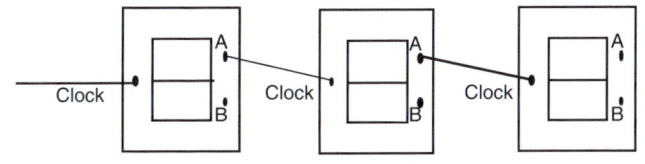

At the start of the count, all the A outputs are LOW and no-one is in the lift. When the doors are shut, the A logic states are 0, 1, 1 (left to right).

13 Flip-flops

Question 564
Which of the following could be the number of people in the lift?
A   2   B   3   C   4   D   5   E   6

Two flip-flops are connected to the output of a square wave oscillator, as shown below. Two of the outputs go to an AND gate, whose output goes to an indicator lamp. The indicator flashes at regular intervals.

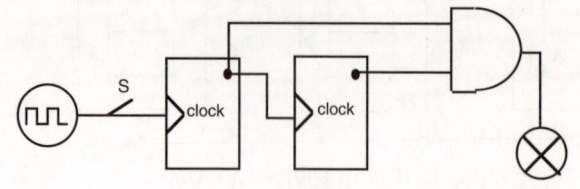

Question 565
Initially, both flip-flop outputs are in logic state 0. They change when the clock input changes from logic state 0 to 1. After the switch S is closed, how many pulses from the oscillator arrive at the clock input of the first flip-flop before the outputs of the flip-flops are again both in logic state 0?

Question 566
After the switch S is closed and three pulses arrive at the first flip-flop, what is the state of the indicator lamp?

Question 567
The oscillator has a frequency of 2 Hz. Its output is graphed below. In the time interval shown on the graph following, state the times when the indicator light is on.

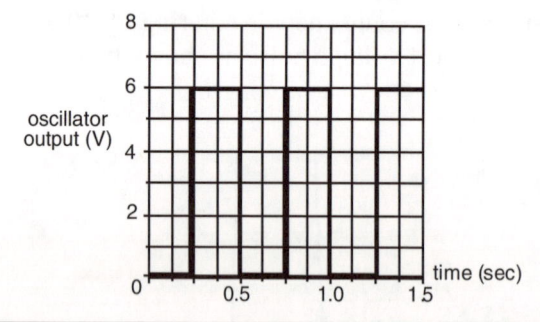

Three flip-flops P, Q and R are connected in series to form a counter, as shown following. Initially all the A outputs are in logic state 0. They change state on a change in the clock input from 0 to 1. Pulses then arrive at the left hand flip-flop every two seconds. The rising edge of the first pulse arrives at time t = 1 seconds. The pulses last for exactly one second.

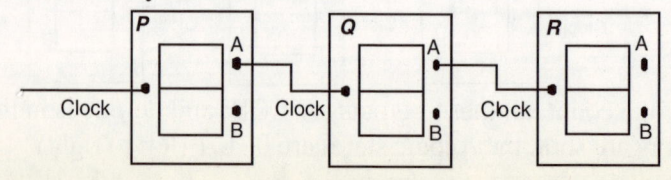

138

13 Flip-flops

Question 568
What time elapses before the A output of flip-flop R changes from logic state 0 to logic state 1?

This counter is now used to turn an appliance off after a certain fixed operating time. The circuit shown on the next page is used.

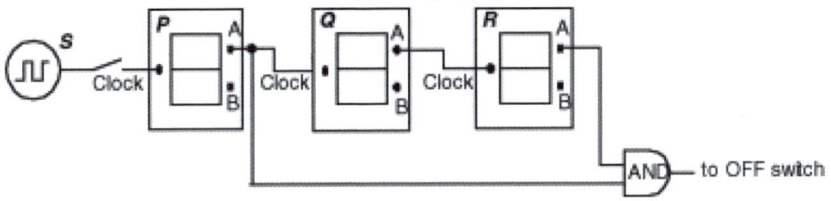

It works as follows.
- All the A outputs are initially at zero.
- The switch S is closed at exactly the same time that the appliance is switched on.
- Pulses feed into the first flip-flop (P) at exactly one per second.
- When the output of the AND gate rises from logic 0 to logic 1 the appliance is switched off.

Question 569
How long does the appliance operate for? Explain your reasoning.

[4 marks]

Pressure sensitive tape can be used to count cars. Each car crossing the tape creates two pulses, one from each of the front and back wheels, as shown. Cars are counted by halving the number of pulses, with a flip-flop. The flip-flop output at Q changes logic on a rising input (0 - 5 V). It remains unchanged on a falling input. The circuit counts car numbers correctly.

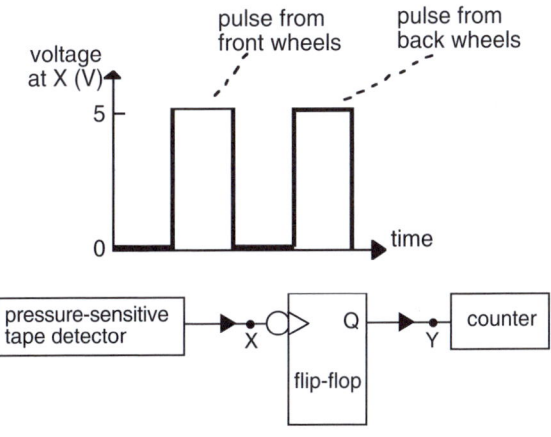

Question 570   (1998 CAT)
Sketch the voltage at Y against time. Indicate how this voltage is related to the voltage at X as a function of time as a car passes over the tape. Assume that the initial voltage at Y is zero.

[3 marks]

139

13 Flip-flops

Trinh sets up a simple digital counter consisting of three flip-flops connected as shown.

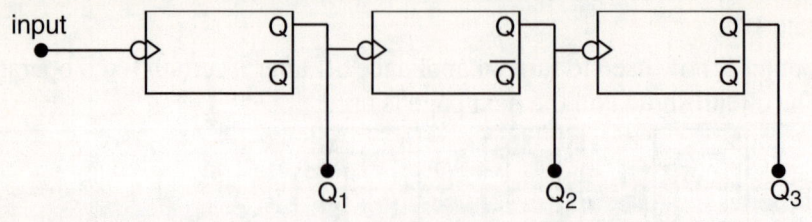

A graph of the input voltage to the counter against time is shown below.

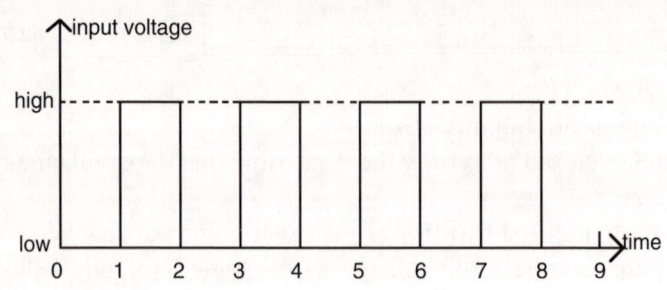

The output of each flip-flop changes state only when its input changes from LOW to HIGH. Initially all flip-flops are in the low state.

Question 571 *(1999 CAT)*
Sketch the output voltage at Q1, Q2 and Q3 for the input voltage shown.

[4 marks]

| Appendix A | Sound summary |

SOUND BASICS
sound is caused by vibrating objects in solids and fluids
- frequency of sound (f) = frequency of vibrating object (measured in hertz Hz)
- period of sound, *T= 1/f;* T is measured in seconds
- different frequency sounds are heard as sounds of different pitch
- high-pitch notes are high-frequency sounds

vibrating objects set up pressure waves in material around them
- this material (solid or liquid) is often referred to as a *medium*
- these pressure waves spread out through the medium, travelling at the speed of sound
- typical values:
    - air   340 m s^{-1}
    - water   1500 m s^{-1}
    - brass   3500 m s^{-1}
- speed of sound depends on temperature
    - as temperature increases, speed increases (generally)
- speed unaffected by pressure or frequency changes

waves transfer energy without transferring matter
- vibrations contain kinetic energy (KE)
- vibrations are 'passed along'; hence so is KE

waves can be transverse
- vibrations are perpendicular to the direction the wave is going

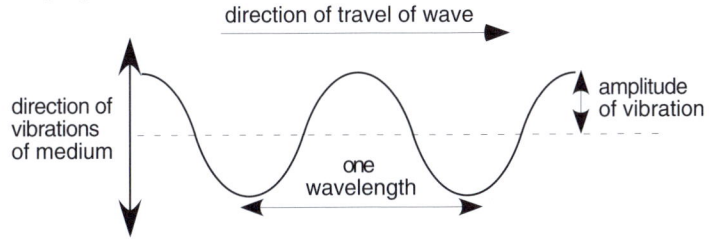

- examples: water waves, 'mexican' waves, waves in ropes and springs
- speed of the vibrations is not the speed of the wave
- *wavelength* is distance between two 'crests' or two 'troughs'
- the *amplitude* is the max displacement of the particles of the medium
- wave moves because adjacent vibrations just out of step (out of phase)

Sound summary

waves can be longitudinal
- vibrations parallel to direction of wave motion; snapshot might look like as follows.

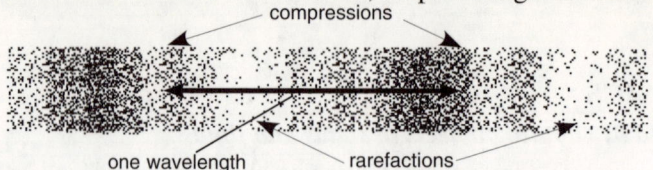

compressions
one wavelength
rarefactions

- examples: sound waves, some waves in springs, some earthquake waves
- speed of the vibrations is *not* the speed of the wave
- wavelength is distance between two compressions or rarefactions
- amplitude is the max. displacement of the particles of the medium
- wave moves because adjacent vibrations are out of step (out of phase)
- in compressions pressure a bit above 'normal' atmospheric pressure
- in rarefactions a bit below 'normal' atmospheric pressure

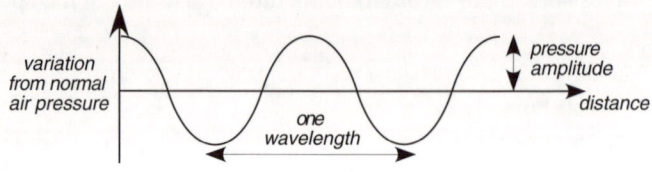

variation from normal air pressure / pressure amplitude / one wavelength / distance

other key terms
- *pressure amplitude (Ap)*: max. variation of pressure from atmospheric pressure
- larger A or Ap, louder sound (more KE being transferred by sound wave)
- *frequency (f)*: no. of vibrations/sec; number of wavelengths passing a point/sec
- *period (T)*: time for one vibration, time for one wavelength to pass a point

key formula: the wave equation: $v = f\lambda$

particle movement in sound waves
- particles are molecules of the medium
- they vibrate parallel to wave direction (hence 'longitudinal')
- they vibrate about a fixed mean position
- they don't 'shuffle along' with the wave

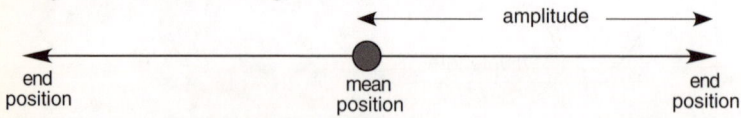

end position — mean position — amplitude — end position

- particles stationary at end positions
- at max. speed at mean position
- in centre of compressions
    - at mean position
    - moving at max. speed
    - moving in same direction as wave
- in centre of rarefactions
    - at mean position particles are moving at maximum speed and in the opposite direction to wave
- in between
    - they stop at the end positions and change direction

Sound summary

frequency response (e.g. of human ear)

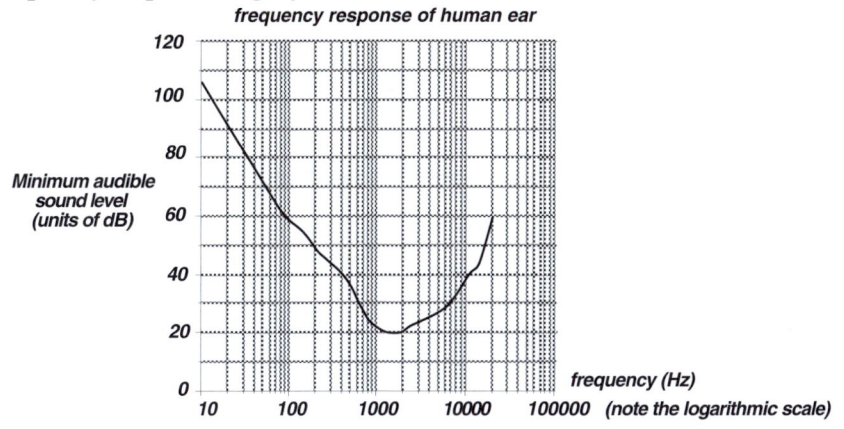

frequency response of human ear

[Graph: Minimum audible sound level (units of dB) on y-axis from 0 to 120, vs frequency (Hz) on x-axis from 10 to 100000 on logarithmic scale. Curve starts high near 100 dB at 10 Hz, decreases to minimum around 1000-10000 Hz, then rises again.]

- shows response of a receiver (e.g. human ear) or transmitter (e.g. loudspeaker)

reflection

- when sound strikes a surface, part of it is absorbed and part is reflected
- absorbing sound energy transfer of KE from air particles to particles of surface
- if surface is hard and smooth, likely to be a good reflector (little energy absorbed)
- reflections occur as shown:

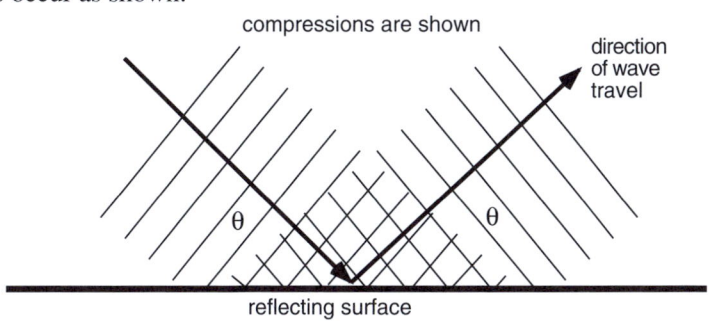

[Diagram showing incident and reflected wave with compressions, angles θ equal on both sides of normal to reflecting surface, direction of wave travel indicated.]

- note that, on reflection:
    - frequency, wavelength and speed unchanged
    - it's not generally 90°, but
    - angle of incidence = angle of reflection (θ)

how sound gets where its going

- must have a medium to vibrate ('in space, no-one can hear you scream')
    - won't travel through a vacuum
- speed depends on medium; speed doesn't depend on f, λ or A

detection

- any device that detects *pressure variations* detects sound
- KE of vibrating molecules of medium is transferred to the detector
    - examples are microphones, ears, dB meters

absorption

- sound which isn't reflected or transmitted is absorbed

Sound summary

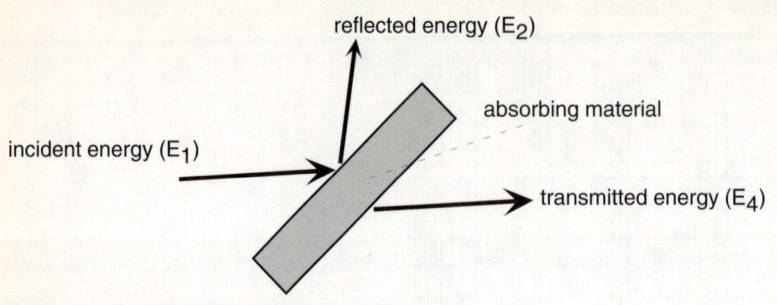

[Diagram: incident energy (E_1) striking absorbing material, with reflected energy (E_2) and transmitted energy (E_4)]

- absorbed energy $E_3 = E_1 - E_2 - E_4$

SOUND INTENSITY AND LEVEL

intensity and loudness (level)
- intensity measures the energy/sec/square metre
- units are W m^{-2} (could be J s^{-1} m^{-2}); symbol is I
- sound intensity *level* measures 'loudness'(sometimes just 'sound level')
    - symbol is L; units are dB (decibels)
    - 0 dB is defined as the sound intensity level of 10^{-12} W m^{-2}
    - connecting formula is: $\mathbf{\mathit{L = 10\ log_{10}I + 120}}$
- common values for L for everyday sounds

sound	intensity (W m^{-2})	level (dB)
threshold	10^{-12}	0
soft whisper	10^{-10}	20
quiet class	10^{-7}	50
traffic	10^{-5}	70
rock concert	1	120

- relative dB levels
- sometimes dB are used to compare two sound levels: $\Delta L = 10 log_{10}[I_1/I_2]$
- some useful values
    - +3 dB means a doubling of intensity
    - -3 dB means a halving of intensity
    - 6 dB means a factor of 4 increase/decrease
    - 9 dB means a factor of 8 change, and so on
    - 10 dB means a factor of 10 change
    - 20 dB means a factor of 100 change, and so on

inverse square law
- as you move away from a sound source, the intensity decreases
- modelled by assuming sound spreads out equally in all directions (not true in practice)
- hence total power of sound source spreads out over a bigger area – this is described by: $I = P/4\pi r^2$, where r is distance from source, P the (total acoustic) power of the source, and I the intensity at a distance r from the source
- this is sometimes written as $I \propto 1/r^2$; described as an *inverse square* law

DIFFRACTION AND INTERFERENCE
diffraction
- when waves pass through an opening, they spread out; this is called *diffraction*
- the amount of diffraction depends on:
    - opening width, d, and wavelength, λ
    - in general, the amount of diffraction $\propto \lambda/d$
    - if $\lambda \ll d$ (say $100\lambda < d$); very little diffraction

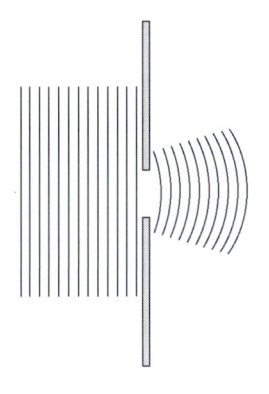

- if $\lambda = d$ (or $\lambda > d$) diffraction is 'complete', as shown below
- waves also diffract around corners
- here d is replaced by the rough size of the object

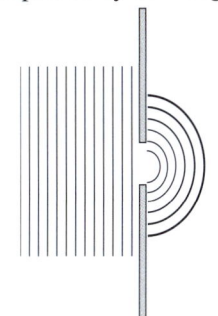

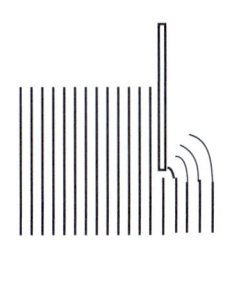

- general points
    - low frequencies and long wavelengths; long wavelengths diffract more than short wavelengths
    - audible sound has λ from ~ 15 m to 2 cm
    - crucial parameter is λ/d
    - loudspeakers: more diffraction from bass notes
    - instruments with small openings diffract sound more (but note that small openings also let out less energy)

superposition
- when two waves cross, their effects just add
- wave 1 and wave 2 just add to 'total'

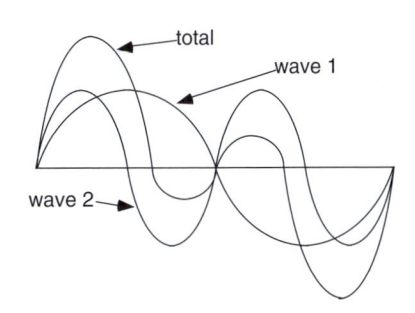

Sound summary

- when two similar waves are 'in phase' (in step) they produce a 'double wave'
- sometimes described as *constructive interference*

- when two similar waves are 'out of phase' (out of step)
- cancellation or destructive interference occurs

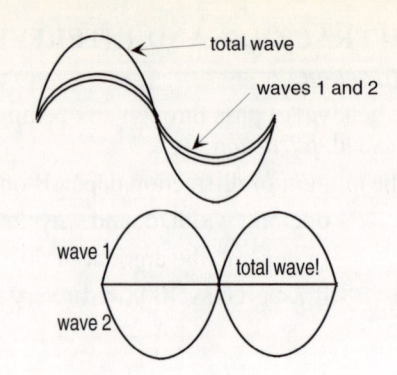

waves can get 'out of step'
- when they start together but travel different distances
- e.g. two loudspeakers producing same frequency (in phase): what happens at P
- when waves arrive at P *in phase*:
    - constructive interference (a loud sound; double the intensity compared to one speaker)
- when waves arrive at P *out of phase*:
    - destructive interference (no sound)

in phase when:	*out of phase* when:
• PY - PX = 0, λ, 2λ, etc.	• PY - PX = $\lambda/2$, 3$\lambda/2$, 5$\lambda/2$, etc.
• PY – PX is the path difference	• PY – PX is the path difference

- special features:
    - perpendicular bisector of XY (dashed line above): always constructive interference, since path difference = zero all along it
    - constructive interference lines are *antinodal* lines (dashed lines)
    - destructive interference lines are *nodal* lines (full lines)

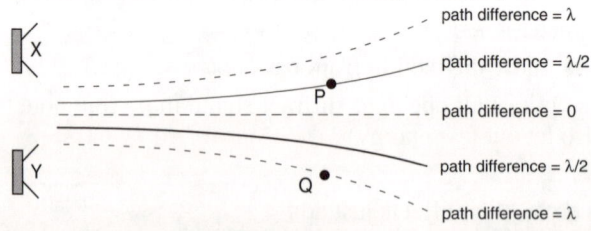

STANDING WAVES AND RESONANCE
all 'mechanical' systems can resonate
- 'mechanical' systems include bridges, beams, buildings, stretched strings, air columns, pendulums, etc.
- they have natural vibration frequencies
- when an external vibration of the same frequency is applied, the system responds with large amplitude vibrations
- other frequencies do not produce large amplitude vibrations
- this phenomenon is called *resonance*

- examples include:
    - singing in the bathroom; certain notes resonate–pushing a swing (only one frequency works)
    - plucking a guitar string; generally only one frequency responds (though you can get harmonics–higher frequencies as well)
    - shaking things (e.g. a tree); some shaking frequencies make it vibrate enthusiastically, others don't (resonance normally means that standing waves have been set up in the mechanical system)
- 'standing' waves don't *travel*
- they have nodes–spots which don't vibrate at all

standing waves are always formed when:
- two waves of the same frequency, wavelength and amplitude travelling in opposite directions overlap
- example: a stretched string (e.g. guitar); possible standing waves shown below
- cause: a wave travels along the string; is reflected at the end and travels back
- original wave and the reflected wave interfere; the result is a standing wave
- simplest (fundamental, or first harmonic) shown below

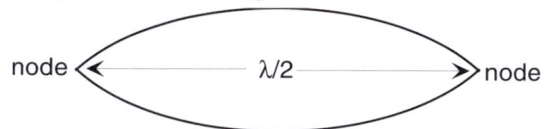

node ⟵ $\lambda/2$ ⟶ node

- the pattern shown is an envelope of the vibrations
- two nodes – one at each end
- distance between nodes is $\lambda/2$
- mid-point is called an *antinode*
- all points in the wave vibrate in phase
- they travel up and down in step
- the wave is standing - doesn't travel left or right

- next simplest is the second harmonic or first overtone node

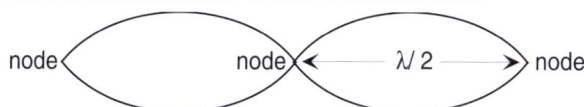

node ⟵ node ⟵ $\lambda/2$ ⟶ node

- within each loop points again travel up and down in phase

- next one is the third harmonic (or second overtone)

node ⟵ node ⟵ node ⟵ $\lambda/2$ ⟶ node

- key formula for wavelengths of these standing waves is: $\lambda_n = 2l/n$
- where l is the length of the string; n = 1, 2, 3, ...

example 2
- two loudspeakers facing each other

A • Q • P B

Sound summary

- the point P will be an *antinode* (waves must be in step, since they have travelled the same distance, therefore constructive interference)
- Q will be a node if BQ − AQ = $\lambda/2, 3\lambda/2, 5\lambda/2, ...$
- Q will be an antinode if BQ − AQ = $0, \lambda, 2\lambda, 3\lambda,....$
- spacing between nodes will still be $\lambda/2$

example 3
- an air column open at both ends (e.g. a flute, didgeridoo, oboe)

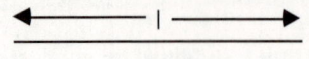

- standing wave of air can be described in *pressure* or *displacement* terms

pressure description	displacement description
• ends open to the atmosphere must be pressure nodes	• ends open to the atmosphere must be displacement antinodes
• nodes at each end	• antinodes at each end
• pressure antinode in centre of tube	• at antinodes displacement variations from normal are maximum
• at nodes pressure variations from normal atmospheric pressure are zero	• displacement node in centre of tube
• simplest standing wave (fundamental, or first harmonic) can be pictured as:	• simplest standing wave (fundamental, or first harmonic) can be pictured as:
• length, *l*, of tube = $\lambda/2$	• length, *l*, of tube = $\lambda/2$
• next simplest standing wave (second harmonic, or first overtone):	• next simplest standing wave (second harmonic, or first overtone):
(three nodes)	(two nodes)

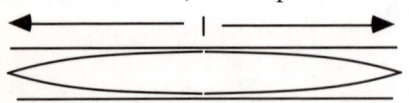

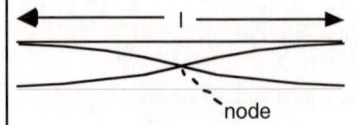

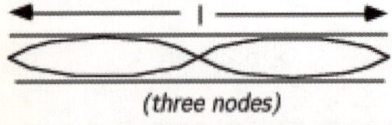

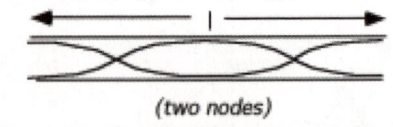

- both descriptions are accurate and equivalent
- key formula for wavelengths of possible standing waves is: $\lambda_n = 2l/n$
- where *l* is the length of the pipe; and n = 1, 2, 3, ...

example 4
- an air column closed at one end (e.g. a clarinet)

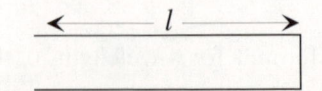

pressure description	displacement description
• end open to the atmosphere must be a pressure node	• end open to the atmosphere must be a displacement antinode
• pressure antinode at closed end	• displacement node at closed end

Sound summary

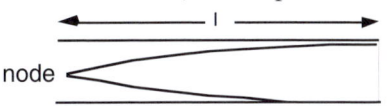

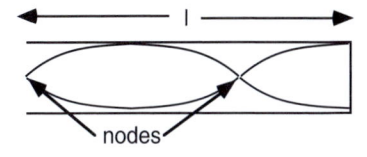

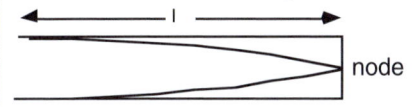

• simplest standing wave (fundamental, or first harmonic) can be pictured as: node ◁≡≡≡≡≡▷ • length, *l*, of tube = $\lambda/4$ • next simplest standing wave (second harmonic, or first overtone): ⟨nodes⟩	• simplest standing wave (fundamental, or first harmonic) can be pictured as: ≡≡≡≡≡▷ node • length, *l*, of tube = $\lambda/4$ • next simplest standing wave (second harmonic, or first overtone): anti-node

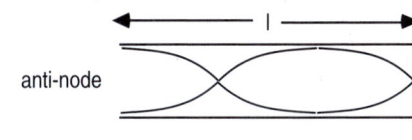

• key formula for the wavelengths of the possible standing waves is: $\lambda_n = 4l/n$
• where *l* is the length of the pipe, and n = 1, 3, 5, ... (odd values only)

pipes with one or two open ends can model
• most wind instruments
  - flutes, recorders, oboes, didgeridoos (two open ends)
  - clarinets (one open end)
• the resonating boxes of stringed instruments
  - guitars, violin body (two closed ends - same formulas as two open ends)
• the ear (one open end)
• the vocal tract (one open end)

stretched strings can model
• most string instruments; the vocal chords

APPENDIX B   Electric power summary

MAGNETIC AND ELECTRIC BASICS
basic electric circuit fundamentals
- current and charge
    - current is the flow of electric charge
    - current is measured in *amperes* (amps, A)
    - charge is measured in *coulombs* (C)
    - symbols: current is I, charge is Q
    - key formula: *I = Q/t*
- voltage and energy
    - voltage is the energy per unit charge; measured in *volts* (V)
    - energy is measured in *joules* (J)
    - key formulas: *V = E/Q, E = VIt*
- power
    - power is the rate of energy transfer
    - symbol is P
    - measured in *watts*, W
    - key formulas: *P = E/t; P = VI*
- Ohm's law
    - conductors obey Ohm's Law if resistance is constant (at constant T) 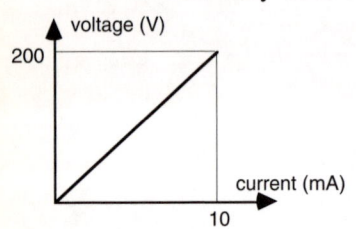
    - resistance is defined by: R = V/I
    - such conductors are called 'ohmic'
    - V vs I graphs look like:
    - the resistance of this conductor: 20 kΩ

resistors in series

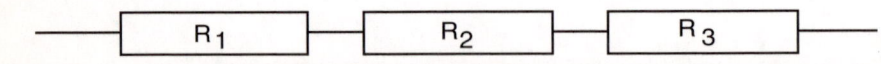

- total resistance is sum: $R_t = R_1 + R_2 + R_3$

resistors in parallel

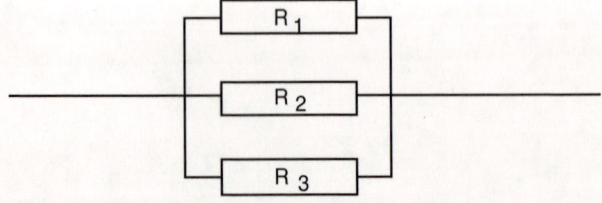

- total resistance is found from: $1/R_t = 1/R_1 + 1/R_2 + 1/R_3$

Electric power summary

magnetic fields from permanent magnets
- out of North, into South

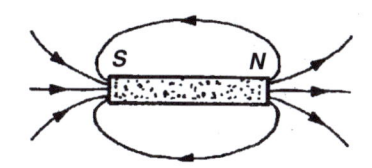

field of the Earth
- magnetic North: a 'physics' South pole

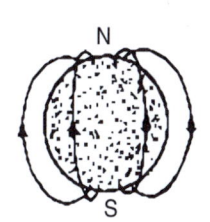

from straight wires
- right hand grip gives direction
- thumb = current; fingers curl in field direction

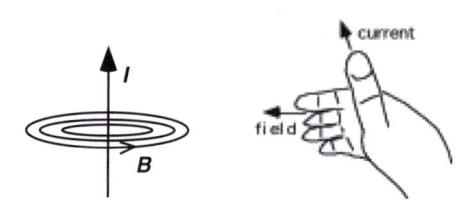

from coils
- use right-hand grip rule to give direction

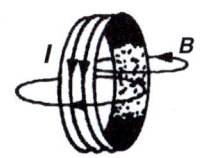

from solenoids
- use right-hand grip rule to give direction

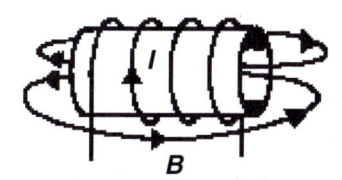

magnetic forces on charges
- only on moving charges; force at right angles to both B and I
- I is parallel to v for *positive* charges; antiparallel for *negative* charges
- tends to push charges in *circular* paths

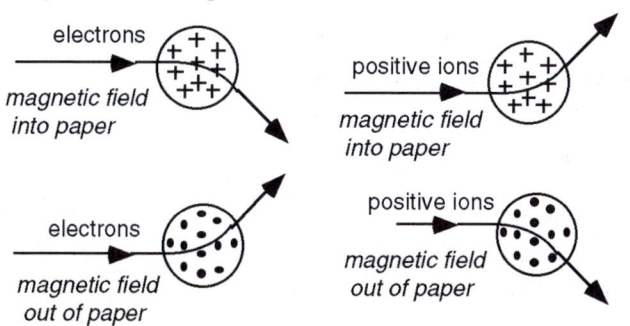

electrons → magnetic field into paper

positive ions → magnetic field into paper

electrons → magnetic field out of paper

positive ions → magnetic field out of paper

151

Electric power summary

magnetic forces on currents
- size of force: F = BIl
- true when B is ⊥ to I
- right hand slap (*RHS*) rule:
- wire is pushed sideways
- if B parallel to I, F = zero
- if B and I at angle θ (0<θ<90), then 0 < F < BIl

AC voltage and current relationships
- V_{peak} (V_p): from zero (axis) to max value of V
- $V_{peak\ to\ peak}$(V_{p-p}): from max to min values of V
- V_{rms} : a sort of 'average'; given by $V_{peak}/\sqrt{2}$
- 50 Hz AC voltage shown (note: current relationships: same as for voltage):

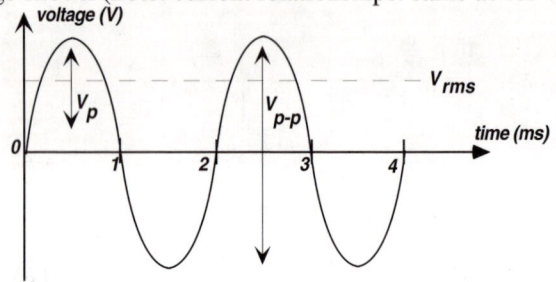

AC power relationships
- average power, $P_{ave} = V_{rms}I_{rms} = P_{rms}$; note this is $P_{ave} = (V_pI_p)/2$
- hence $P_{peak} = V_pI_p = 2P_{rms}$

frequency of AC (f)
- no. of times cycle repeats per second; measured in Hertz (Hz)
- normal mains supply is 50 Hz; period of AC (T) = 2 ms (time for one cycle)
- *f = 1/T; T = 1/f*

MOTORS
current-carrying coils in B fields
- forces act as shown above (RHS rule)
- this causes a torque (twist effect)
- end-on view is easier to draw, as shown:
- pushes coil to vertical position
- here the current is disconnected momentarily
- momentum carries it past the vertical position
- current then connected in opposite direction

- this is achieved by the *split-ring commutator*
- torque thus pushes coil in the same direction as before
- hence current is reversed in direction every half turn
- torque is maximum when coil is parallel to B
- torque is zero when coil is perpendicular to B

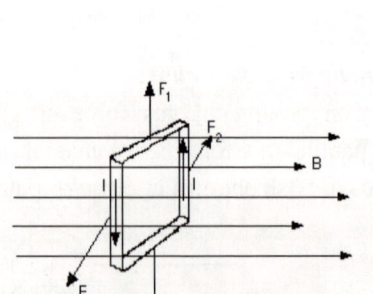

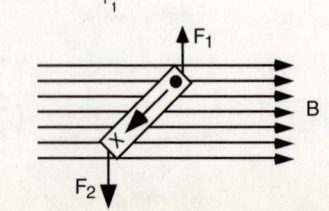

152

Electric power summary

- size of max. torque $\propto$ nBAI, where:
  - n = number of turns
  - B = strength of magnetic field
  - A = area of coil
  - I = current through coil

split-ring commutator detail (simple treatment)

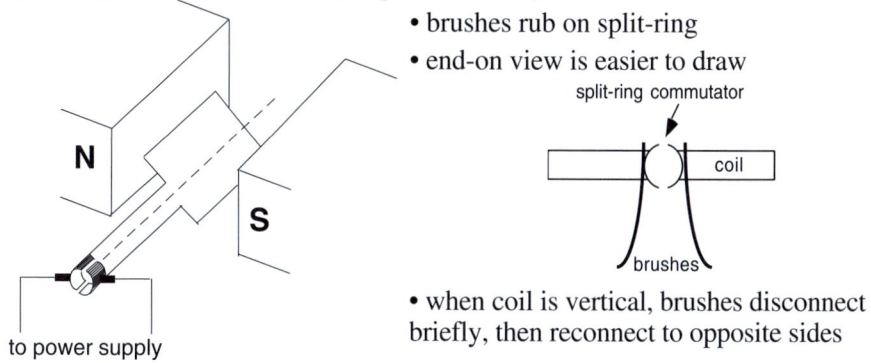

- brushes rub on split-ring
- end-on view is easier to draw

split-ring commutator / coil / brushes

to power supply

- when coil is vertical, brushes disconnect briefly, then reconnect to opposite sides

alternative way of looking at motors

- rotating solenoids in magnetic fields

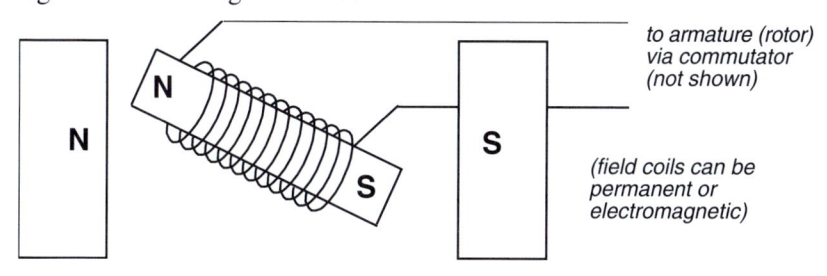

to armature (rotor) via commutator (not shown)

(field coils can be permanent or electromagnetic)

- this form is best analysed by looking at *poles* of rotating solenoid
- commutator switches current direction every half turn
- this reverses poles of armature, ensuring continued rotation in the same direction (as before)
- switching happens when axis of solenoid is *parallel* to B (compare with flat coil motor previously)
- maximum torque when axis is perpendicular to B
- for these motors torque $\propto$ nBI
- n is no. of turns per unit length (in rotor); B is strength of magnetic field; I is current in armature

AC motors

- rotate at the same frequency as the AC
- use the *alternation of the current* instead of a commutator for the armature
- slip-rings connect current to the armature (cf with split-ring commutator)

153

Electric power summary

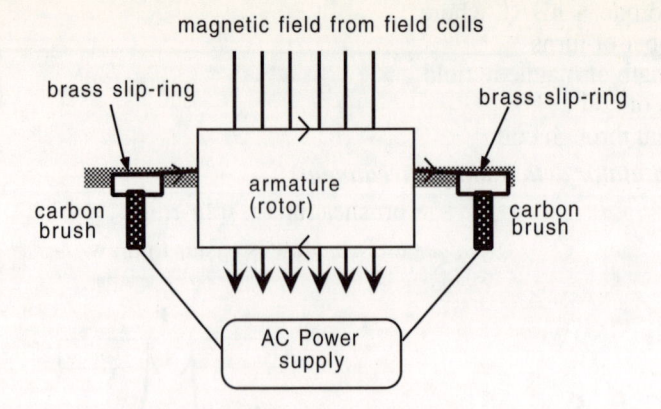

- note that this kind relies on a DC field from the stator
- if both stator and rotor run off AC a commutator is needed; in this case the motor 'mimics' a DC motor

GENERATION PRINCIPLES

emf (voltage) induced when wires 'cut' magnetic field lines

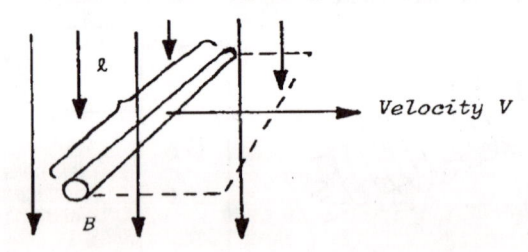

- emf produced given by $\varepsilon = Blv$ when wire cuts magnetic field lines
- ε = emf (V); B = magnetic field strength (T); l = length of wire (m); v = speed of wire (m s^{-1})
- this formula is only accurate when v is perpendicular to B
- when v is parallel to B, ε is zero; at other angles, value is in-between

example

- mast cuts B
- emf = Blv
- no *current*, no closed circuit
- only emf produced

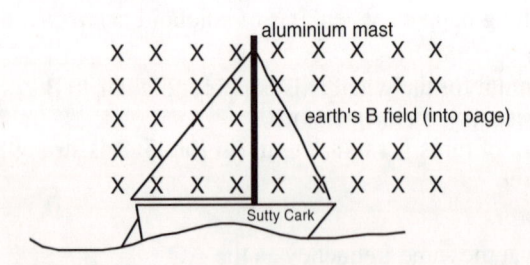

154

magnetic field changing through coils
- when the magnetic field through a coil changes, emf is always generated
- this can be caused by
    - rotating coil
    - changing coil area
    - changing B strength
    - using 'AC' field

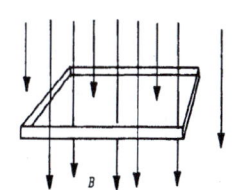

- Faraday's Law gives $\varepsilon = -n\, \Delta(BA)/\Delta t$
    - this is the key equation
    - n = number of turns in coil
    - BA = magnetic flux (true when B is $\perp$ to A) = Φ
    - when B is not $\perp$ to A, flux is less
    - when B is parallel to A, flux is *zero*
    - hence $\varepsilon = -n\, \Delta\Phi/\Delta t$
    - units of flux are webers (Wb)
    - t = time (measured in secs)
    - minus sign indicates direction (see later)
- note that emf is *only* produced when the flux is changing

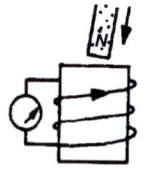

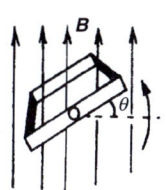

using flux-time graphs
- if a graph of Φ vs t is available, gradient ($\times$ n) gives the emf; e.g.

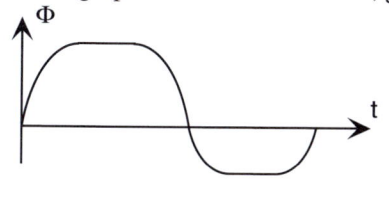

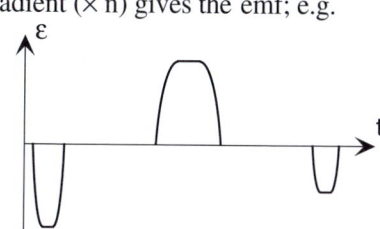

coil rotating in B field
- changing flux through coil causes emf; hence a generator is made; a DC motor in 'reverse' will act as a generator
- flux through coil as shown in left hand graph following

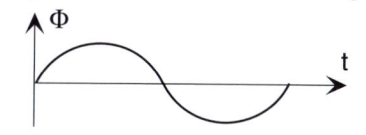

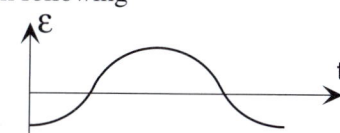

- emf generated as shown in RH graph above
- note when gradient of $\Phi > 0$, then $\varepsilon < 0$ (see later)
- note that the induced emf in the coil is AC
- commutator needed to transform AC to DC: graph following *after* commutator:

Electric power summary

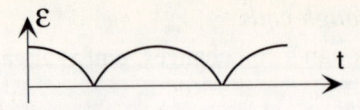

- max emf when coil is parallel with B
- zero emf when coil is perpendicular to B
- peak (max) value of emf $\varepsilon \propto nBAf$
- n = no. of turns; B = magnetic field through coil; A = area of coil; f = frequency of rotation of coil
  - increasing frequency does two things
  - increases frequency of the generated AC, *and*
  - increases peak value of the generated emf
- **direction of induced emf from Lenz's Law; gives direction of induced emf**
- direction *opposes the change* that caused it
- if increasing flux causes the emf, induced current will try and create an *opposing* flux
- if decreasing flux causes the emf, induced current will try and create a *supporting* flux
- if movement causes the emf, there will be 'magnetic drag' (a BIl force) in the other direction
- hence the minus sign in: $\varepsilon = -n\,\Delta\Phi/\Delta t$ and $\varepsilon = -Blv$

AC generators

- these can be viewed as AC motors (see earlier) working 'in reverse'
- coil rotating in steady (DC) field produces AC induced emf
- fed out using *slip rings* rather than a *split-ring commutator*
- often described as *alternators*
- as for DC generators, max emf when coil plane is parallel with B, minimum emf when coil plane is perpendicular with B
- generators with solenoid type rotors analysed in a similar way

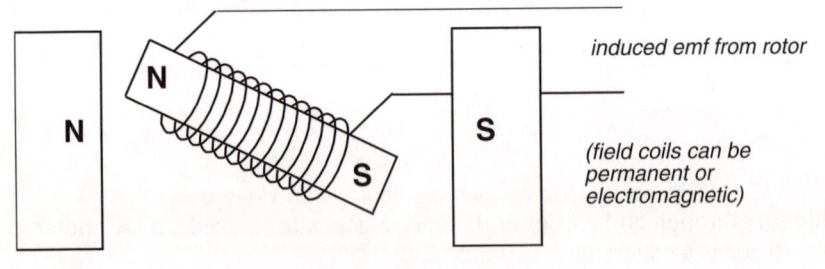

induced emf from rotor

(field coils can be permanent or electromagnetic)

TRANSFORMERS, TRANSMISSION AND CONSUMPTION

basic structure

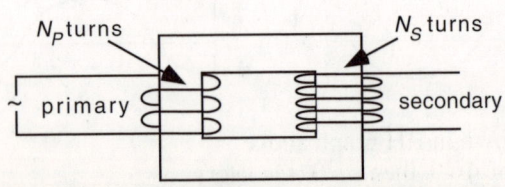

N_P turns — primary
N_S turns — secondary

- primary coil produces magnetic flux; this is an 'AC' flux
- core strengthens and channels the flux through to the secondary coil

Electric power summary

- in an ideal transformer, 100% of the flux produced by the primary is transferred to the secondary
- in the secondary coil, the changing flux produces an induced emf
- when the current in the primary is (sinusoidal) AC, the induced emf in the secondary is also AC
- key formula: $V_P/V_S = N_P/N_S$
- often also true that: power into primary = power out from secondary; this leads to second key formula: $I_P V_P = I_S V_S$
- hence 'step up' transformers also step down the *current* (and vice-versa)

transmission of electricity

- involves voltage losses; key formula is $V_{loss} = IR$, where V_{loss} is voltage loss in the transmission cables; I is the current in cables, R is resistance of cables
- involves power losses
- $P_{loss} = I^2 R$ (I and R as above)
- to minimise losses, reduce R and I
- R can be reduced by thicker cables (or shorter cables – not always practical)
- R can also be reduced by using special materials, e.g. copper and aluminium (good conductors but expensive; superconductors not yet practicable
- often I is reduced by using step-up transformers

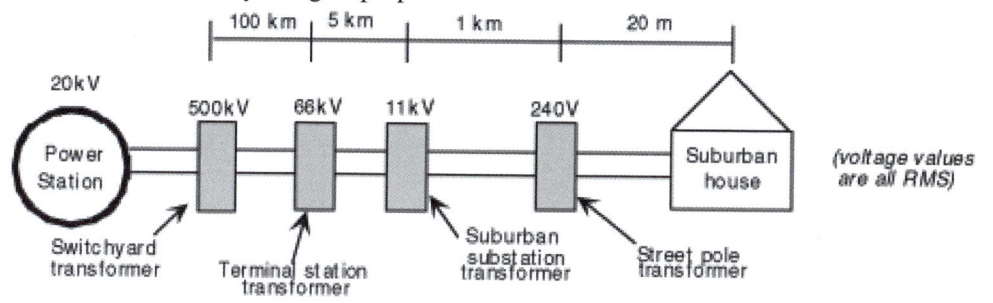

Power Station — 20kV — Switchyard transformer — 500kV — (100 km) — Terminal station transformer — 66kV — (5 km) — 11kV — (1 km) — Suburban substation transformer — 240V — (20 m) — Street pole transformer — Suburban house

(voltage values are all RMS)

domestic electricity costs

- paying for it
    - usually billed in kW hr (*not* kW/hr or kW hr^{-1})
    - these are units of *energy*)
    - 1 kW hr = 3 600 00 J

connection
    - as 240 V (rms) AC, 50 Hz
    - supplied via active and neutral lines
    - active is at 240 V (AC), neutral is at 0 V
    - earth wire is connected to a water pipe; this makes good electrical contact with earth, and ensures earth lines are held at 0 V

principles of switches, fuses etc
    - always switch in the active line
    - sometimes switch in both (for safety)
    - fuses likewise in active line

157

Electric power summary

domestic appliances – earthed kinds

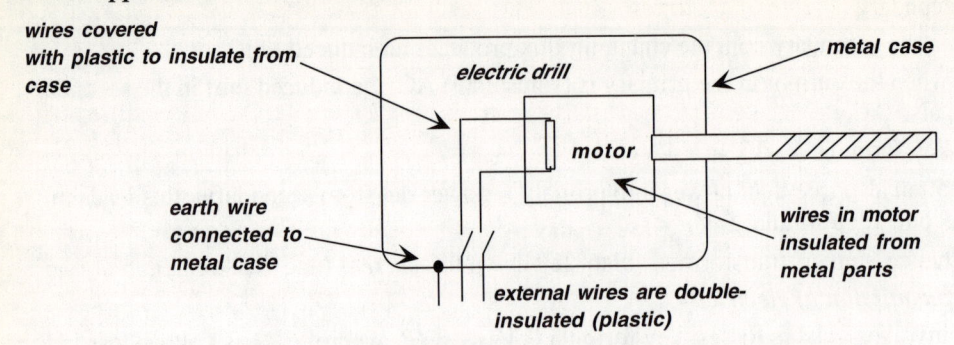

- role of earth wire: connected to case of appliance; provides low resistance path to 0 volts (in case active wire comes into contact with case); when this happens generally a large current drawn along this low-resistance pathway; resulting in a 'blown' fuse

domestic appliances – double insulated

- all active wires inside and outside case are covered with *two* layers of insulation
- case is often plastic (insulator); not now needed to carry away 'leaking' current

earth leakage protection

- often described as RCI (residual current interrupt); one version is sketched following, in a domestic situation

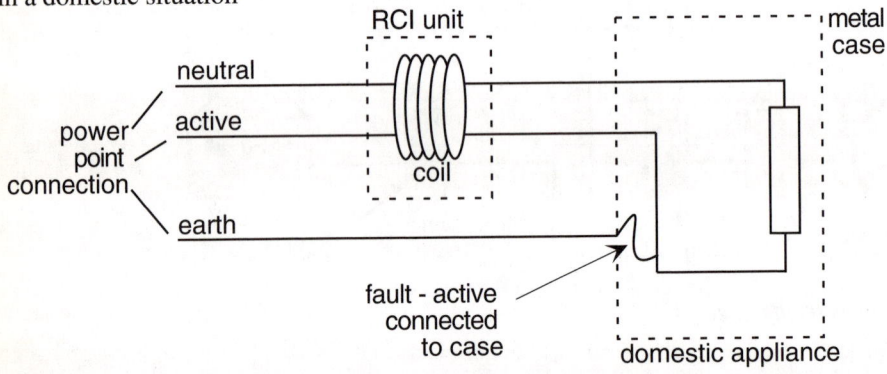

- when currents in the active and neutral lines are equal, there is no changing flux through the coil in the RCI unit
- when they are *un*equal - for example when current 'leaks' to earth via the case – changing flux through coil induces currents in the coil; which can be used to turn off the electricity.

electrical safety

- danger occurs when >100 mA flows through your heart
- from Ohm's Law $I = V/R$; hence safety requires high R and/or low V
- low R means dry hands, rubber soled footwear, etc.

energy usage and load curves

- sample load curve is shown following

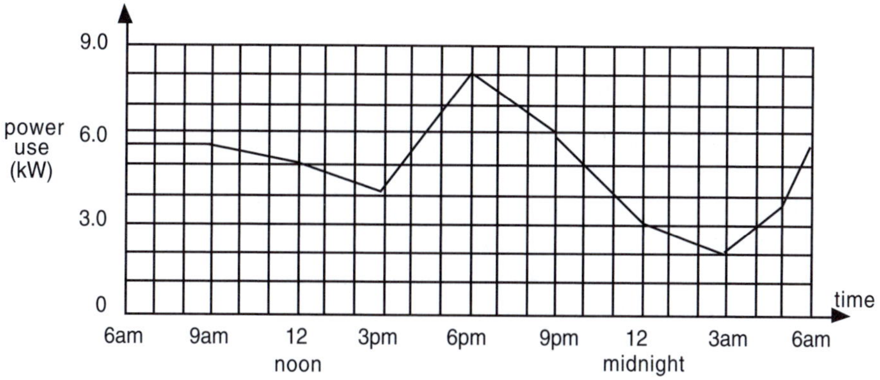
9.0
power use (kW)
6.0
3.0
0
time
6am
9am
12 noon
3pm
6pm
9pm
12 midnight
3am
6am

- area under this curve is equal to the energy consumption for the time period
- take care with the units – in the above graph each square is 1 kW h (3.6 MJ)

APPENDIX C	**Electronics summary**

ELECTRONICS BASICS
(many ideas covered in Electric Power Basics – see earlier)
non-ohmic resistors
- devices with non-constant resistance
- example: torch globe; typical V-I graph follows:

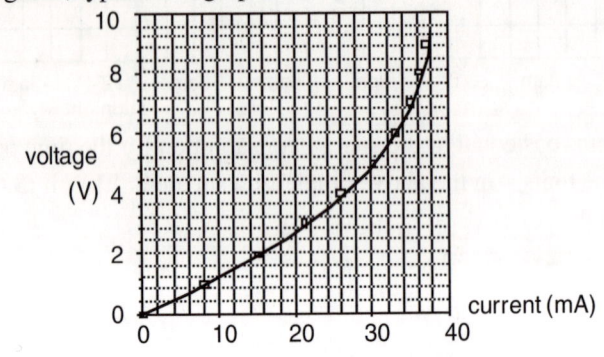

- definition of resistance remains as a simple ratio: **$R = V/I$**

diodes
- only allow current to flow in one direction; typical V – I graph follows:

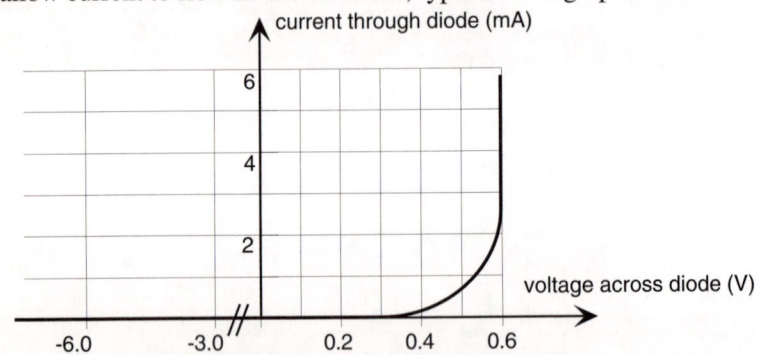

- in forward direction, resistance varies considerably
- in reverse direction ('reverse bias') resistance is very high indeed
- symbols for two kinds of diode shown below:

                'ordinary' diode   *light-emitting diode*

- direction of arrow shows 'forward' direction

Electronics summary

light dependent resistors (LDRs)
- typical graph follows

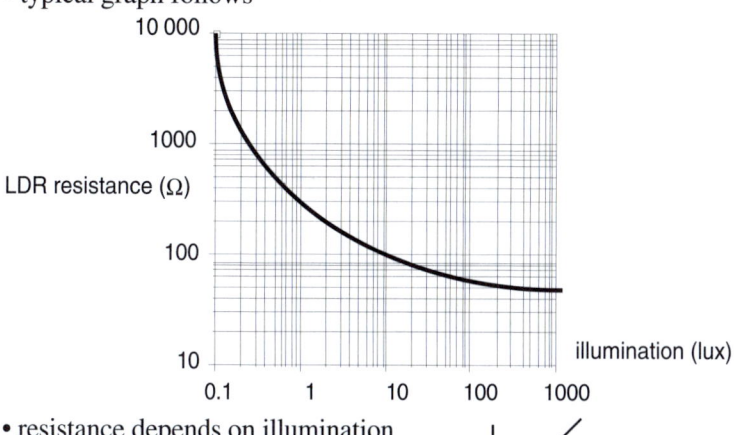

- resistance depends on illumination
- high light levels – low resistance
- low light levels – high resistance
- symbol is shown to the right
- useful as input transducer to gates

thermistor
- resistance varies with temperature; normally falls with increasing T
- typical V – I graph shown below, also symbol

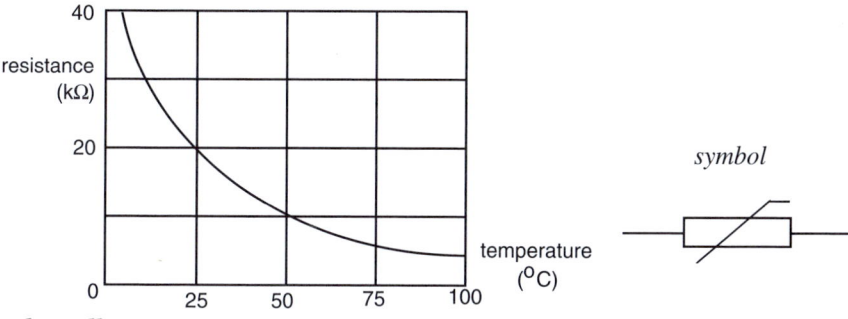

symbol

solar cells
- convert solar radiant energy to electricity; typical characteristic shown

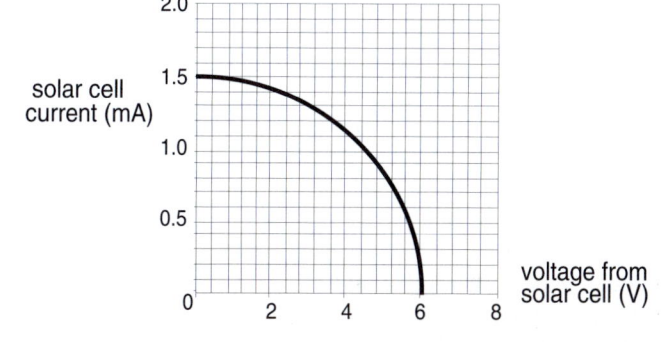

161

Electronics summary

- note that low current means high voltage, low voltage means high current
- max power is generated somewhere in the middle; at this point VI is max.

voltage dividers...
- two resistors arranged as follows:
- V_{out} is a *fraction* of V_0
- resistors R_1 and R_2 'divide up' the voltage V_0 in proportion 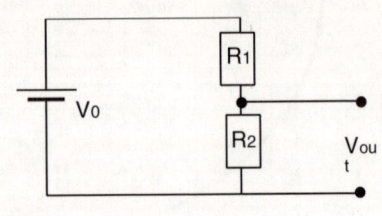
- key formula is $V_{out} = \dfrac{R_2}{R_1 + R_2} V_0$
- V_{out} can be varied if one resistor is variable, as shown below: 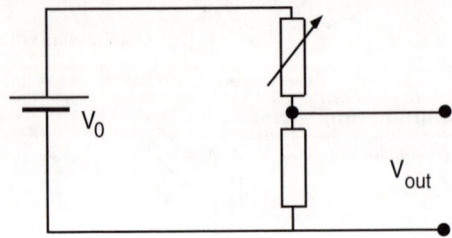
- such units can be used as inputs, using transducers as the variable component

example: convert light level to a voltage
- as light level changes, LDR resistance changes
- at high light levels, LDR resistance is low, hence voltage at X is LOW
- at low light levels, voltage at X is HIGH
- variable resistor used to adjust 'switching' light level

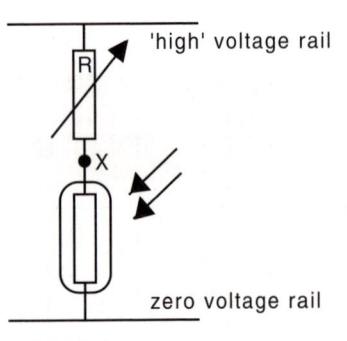

example: temperature to a voltage

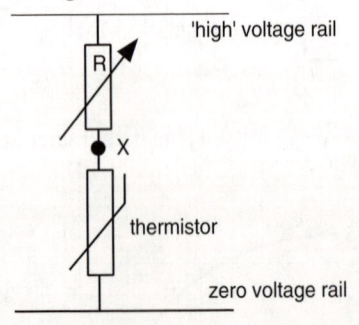

input/output transducers
- microphone (input)
    - symbol
    - converts sound waves into AC signals
    - often used at the input of an amplifier
    - voltages generated are small (mV or less)

Electronics summary

- mechanical switches (input)
    - symbol
    - connect to LOW or HIGH voltage states
    - normally connect to a voltage rail
- loudspeaker (output)
    - symbol
    - convert AC signals to pressure variations; hence sound waves

meters
- measure voltage or current
    - ammeters in series; voltmeters in parallel
- cathode ray oscilloscopes
    - can be used as a voltmeter, with the timebase *off*

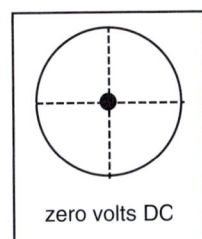

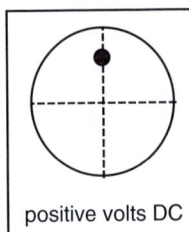

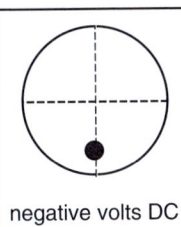

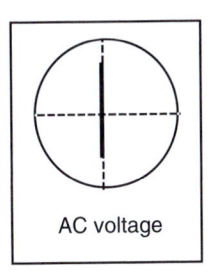

zero volts DC	positive volts DC	negative volts DC	AC voltage

- spot on screen moves *up* for positive voltages; *down* for negative voltages
- spot draws a vertical line for AC voltages; length of this line gives Vp-p

oscillators
- these are electronic devices to produce *repeated* AC voltage signals, as shown

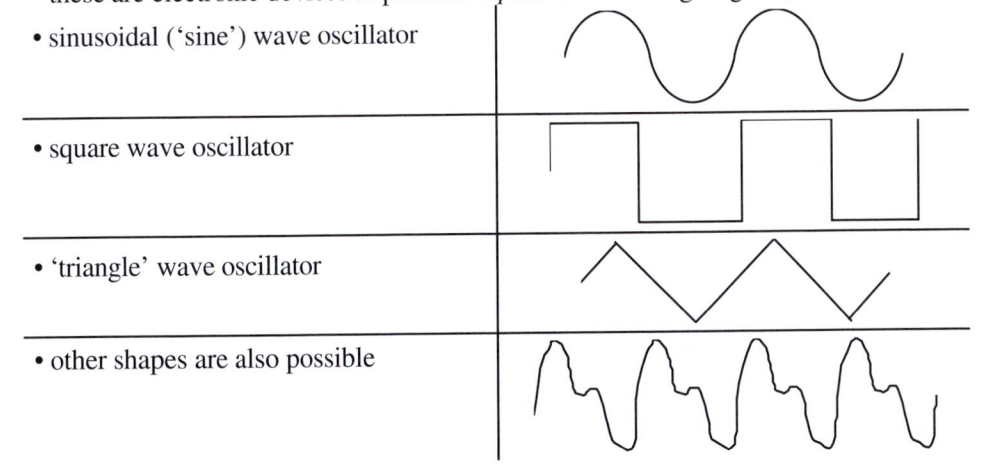

• sinusoidal ('sine') wave oscillator	
• square wave oscillator	
• 'triangle' wave oscillator	
• other shapes are also possible	

163

Electronics summary

CAPACITORS AND DIODES
rectification...
- simple half-wave rectification uses one diode as shown following:

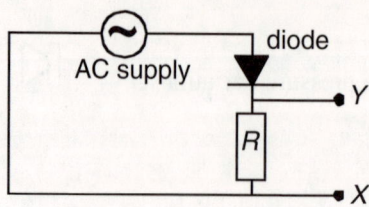

- input is AC: output is 'lumpy' DC (as shown following)

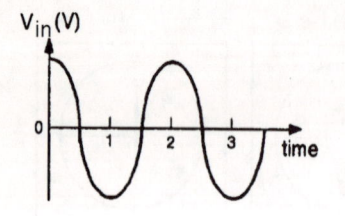

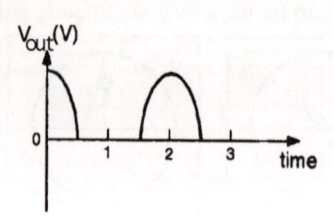

- diode simply prevents negative current
- average value of voltage is now > 0
- Y will be at a higher potential (voltage) than X

capacitors
- these are devices for storing charge
- normally constructed from closely spaced parallel conductors
    - sometimes rolled up in a little cylinder
- symbol C, circuit symbol shown
- units are farads (F)
- normally microfarads (µF); sometimes nF or pF
- amount of charge stored (Q) given by $Q = CV$
    - Q is the charge in coulombs,
    - C is the capacitance in farads,
    - V is the applied voltage in volts.
- example:
    - a 100 µF capacitor charged from a 20 V battery
    - stores Q = CV = 0.0001 × 20 C = 0.002 C = 2 mC.

Electronics summary

charging capacitors
- charging circuit shown; also charge-time graph

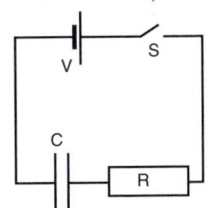

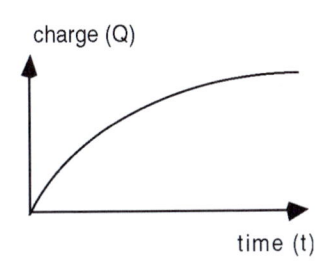

- when S closed, with C uncharged, charge flows from battery on to the plates of the capacitor; when plates are 'full', charge stops flowing
- charge flows fast at first, then less fast, asymptotes to max. charge
- voltage and current time variation shown following

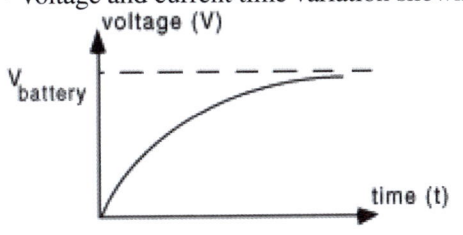

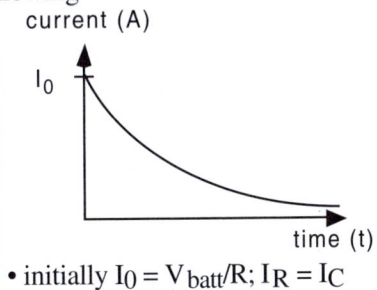

- shape as for charge graph (from Q = CV)
- voltage asymptotes to battery voltage

- initially $I_0 = V_{batt}/R$; $I_R = I_C$
- current asymptotes to 0: $I_R = I_C$

- voltage across resistor
- since V = IR, shape is same as current graph
- initial voltage is $V_{battery}$

- total voltage = $V_R + V_C = V_{battery}$

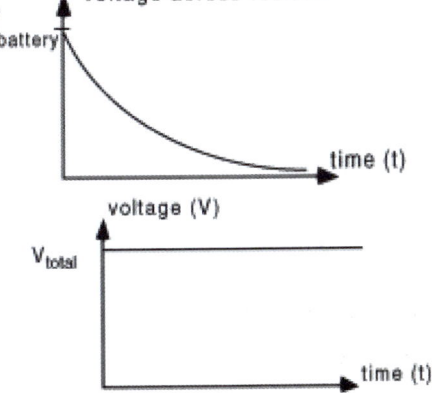

discharging capacitors
- discharging circuit shown

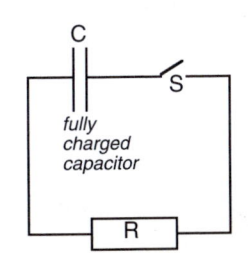

fully charged capacitor

165

Electronics summary

- since $V_C = V_R (= IR)$ in this circuit, both current and voltage graphs have the same shape

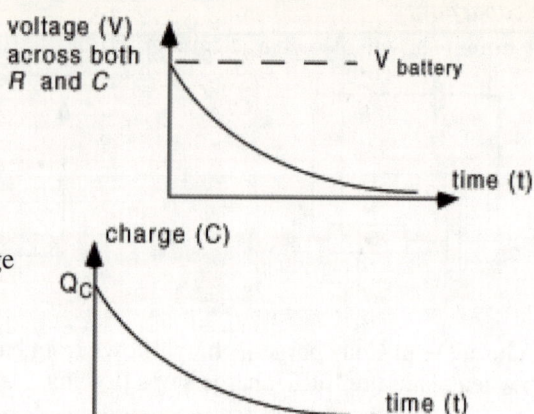

- capacitor discharge graph the same shape as voltage discharge graphs since $Q = CV$

rate of charge and discharge
- rate of charging and discharging is governed by the product **RC**
- this is known as the *time constant* (τ)
- measured in seconds; farads × ohms = secs
- example: RC for 80 kΩ and 35 µF: τ = 80 000 x 0.000035 = 2.8 s.
- from the graph, the current/charge/voltage drops by a factor of e^{-1} in *one time constant*

smoothing
- half-wave rectified DC (see earlier) is too 'lumpy' for many uses; hence requires 'smoothing'
- this is done by connecting a capacitor in parallel with the output, as shown.

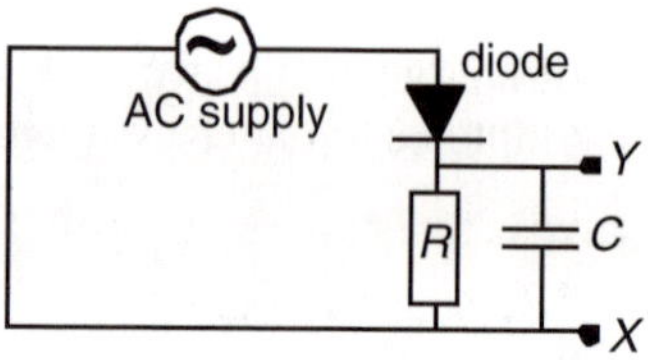

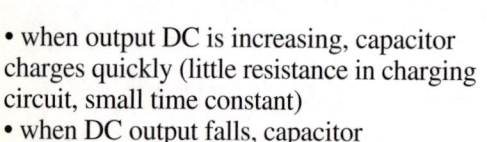

- when output DC is increasing, capacitor charges quickly (little resistance in charging circuit, small time constant)
- when DC output falls, capacitor discharges through resistor R
- this is slow, since governed by time constant RC, result as shown
- larger RC – smoother waveform
- note that RC must be much bigger than the 'gaps' for effective smoothing 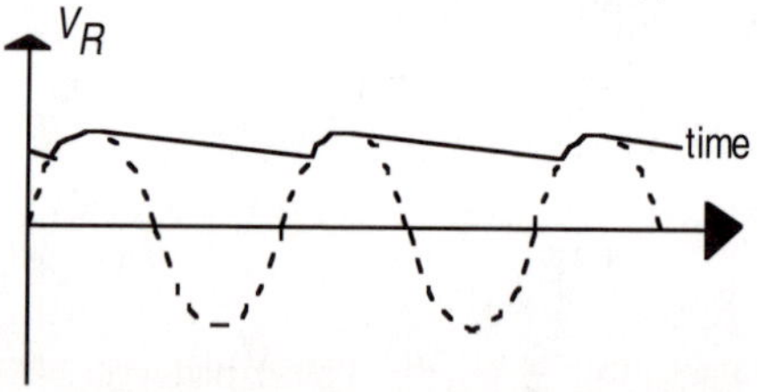
- for 50 Hz AC, t >> 10 ms

AMPLIFICATION
voltage amplifiers simply amplify AC voltage signals

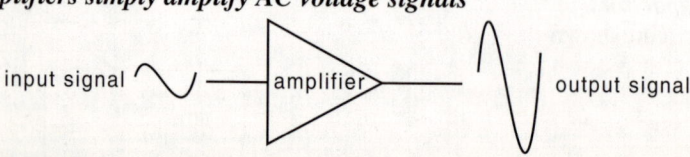

- the ratio $\dfrac{\text{output } V_{p\text{-}p}}{\text{input } V_{p\text{-}p}}$ is the *gain*

DC characteristics
- a typical DC input/output graph is shown below

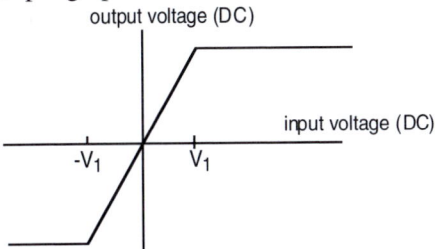

output voltage (DC) / *input voltage (DC)*, with breakpoints at $-V_1$ and V_1

- the gradient of the graph between $-V_1$ and $+V_1$ is the *gain* of the amplifier
- this is a *non-inverting amplifier*
- *inverting* amplifiers have characteristics as shown (gradient still gives the gain)

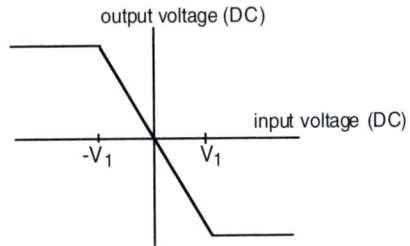

output voltage (DC) / *input voltage (DC)*, with breakpoints at $-V_1$ and V_1

- difference: output AC *out of phase* with input (gain expressed as $-^{ve}$)
- comparison with non-inverting (i.e. normal) amplifier is shown following.

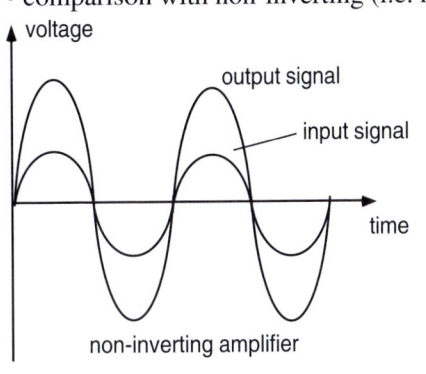

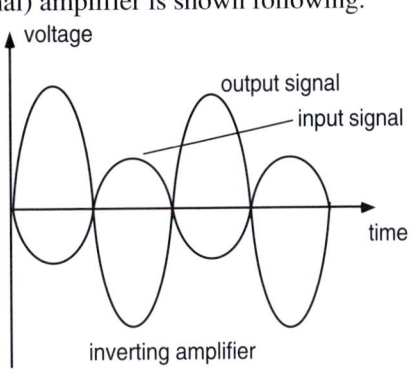

non-inverting amplifier | inverting amplifier

LOGIC
gate logic

name	symbol	truth table
NOT	(triangle with bubble)	IN / OUT: 0→1, 1→0
AND	A, B → O	A B O: 0 0 0; 0 1 0; 1 0 0; 1 1 1

Electronics summary

NAND
(equivalent to an AND followed by a NOT)

A	B	O
0	0	1
0	1	1
1	0	1
1	1	0

NOR

A	B	O
0	0	1
0	1	0
1	0	0
1	1	0

OR
(equivalent to a NOR followed by a NOT)

A	B	O
0	0	0
0	1	1
1	0	1
1	1	1

combinations of gates
• example

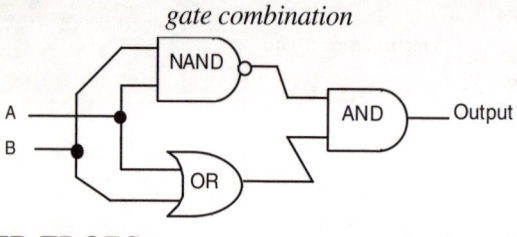

gate combination

truth table

A	B	O
0	0	0
0	1	1
1	0	1
1	1	0

FLIP-FLOPS
flip-flop logic

• note two outputs; Q and $\bar{Q}$
• symbol as shown
• outputs are never the same; one will be HIGH and the other LOW
• clock input *changes* current state
• will change on a *rising* pulse; i.e. input changing from LOW to HIGH
• sometimes called *bistables*

uses of flip-flops
• to lengthen pulses (double length)
   - one pulse per second in
   - one leading (or falling) edge per second in
   - one change in the output per second
   - this means one pulse out every two seconds
   - if input pulses look like:

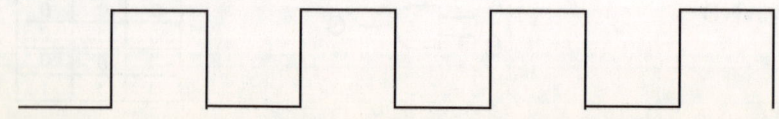

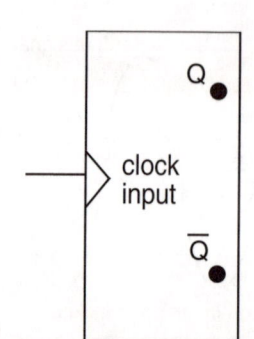

Electronics summary

- then output pulses look like:

- example: if pulse input frequency is 380 Hz, then pulse output frequency will be 190 Hz
- to form a *counter*, using a series of flip-flops as shown

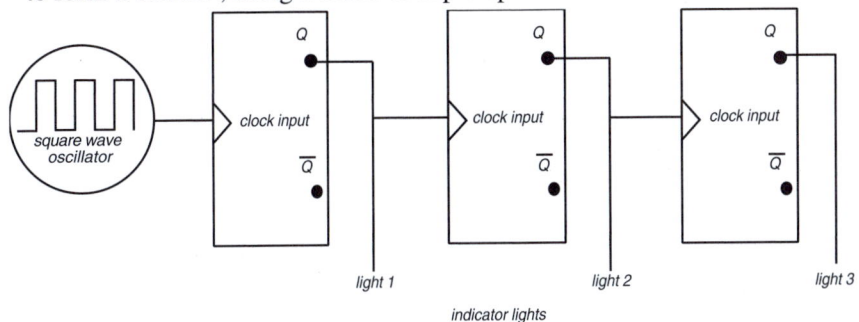

indicator lights

- if the square wave oscillator produces pulses at 1 Hz, then the first indicator flashes at 2 Hz; the next two flash at 4 Hz and 8 Hz.
- when light 1 rises, light 2 changes
- when light 2 rises, light 3 changes
- light 1 changes every pulse
- light 2 changes every second pulse
- light 3 changes every fourth pulse
- the patterns of on and off indicators give 8 different patterns, as shown
- hence this arrangement can 'count' up to 8
- these patterns can be used to stand for numbers
- 0,0,0 stands for 0; 0,0,1 stands for 4; 1,0,1 stands for 6
- such outputs can be used to control electronic systems, often through logic gates

light 1	light 2	light 3
0	0	0
1	1	1
0	1	1
1	0	1
0	0	1
1	1	0
0	1	0
1	0	0

ANSWERS

Sound basics

1. B — The wavelength is the distance between adjacent equivalent points on the wavetrain.
2. 34 cm — The distance between adjacent compressions.
3. 1.0 kHz — $f = v/\lambda = 340/.34$
4. AD — The pressure is greatest where the molecules are most closely spaced.
5. • the molecules in a sound wave vibrate in a direction parallel to the direction of travel of the wave (this is known as *longitudinal* vibration)
   • in a transverse wave, the vibrations are at right angles to the direction of travel of the wave (this is known as transverse vibration)
6.

variation from atmospheric pressure

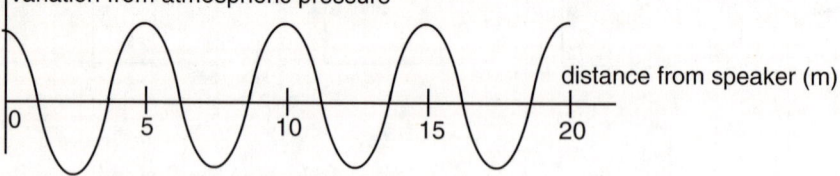

(graph showing sinusoidal wave with distance from speaker (m) on x-axis, marked at 0, 5, 10, 15, 20)

7. $\lambda = 5$ m — read wavelength from graph
   $f = 66$ Hz — use $v = f\lambda$
8. 200 Hz — use $f = 1/T$
9. 1000 — Inverse ratio to frequencies.
10. C — The example of waves passing through would be two people speaking to each other at the same time. An example of a combining of effects might be the standing wave or interference pattern set up using two speakers.
11. 250 Hz — $f = 1/T$; $T = 0.004$ s (from graph)
12. 1.38 m — from $v = f\lambda$
13.

X ▓▓▓▓▓▓▓▓▓▓▓▓▓▓▓▓▓▓▓▓▓▓▓▓ Y

The travelling wave will have moved half a wavelength along the pipe.

14. 65.6 cm — $\lambda = v/f = 336/512$
15. 8.92 ms — $t = d/v = 3/336$ s.
16. • speaker cone vibrates regularly, causing alternate compressions and rarefactions
    • these propagate as a sound wave
    • sound is a longitudinal wave – individual regions of air oscillate back and forth at a frequency of 512 Hz along a line parallel to the sound wave
    • this motion is in addition to the chaotic motion of individual molecules
    • the energy is transmitted by molecular collisions
17. Sound from the band is heard directly, but also reflected from the wall (the incident and reflected angles are equal) which acts like a mirror does for light, causing an 'image' of the band behind the wall.
18. 680 m — must be double 340 m, because of the laws of reflection for sound (incident angle = reflected angle)
19. 339 m — equal to $\sqrt{(480^2 - 340^2)}$
20. C — The speed of the sound wave would be greater in the higher temperature soil. The path would be unchanged.

Answers

21	0.625 s	T = 1/f
22	384 m s⁻¹	v = d/t = 240/0.63
23	She hears the echoes midway through the interval between each clap.	
24	She hears an echo coinciding with each clap, each echo being caused by two claps previous.	
25	A E	The sound would reach her more quickly. She needs to clap faster or move further back to increase the echo interval to its original value.
26	Pulse 1 is the direct drum sound. Pulse 2 represents an echo from the wall. It will have smaller amplitude because of absorption and also the greater distance to travel. The time t represents the extra travel time of the reflected wave compared to the direct wave.	
27	500 Hz	Use f = 1/T (period)
28	350 m s⁻¹	Use v = fλ
29	*(graph showing pressure vs distance, sinusoidal wave about normal atmospheric pressure)*	
30	0.70 m	one wavelength
31	B	• the molecules do not move along the tube (only the wave does) • one period later the pattern of compressions and rarefactions will be the same
32	*(diagram of sloping surface reflecting sound)*	The sloping surface tends to reflect the sound into the sky, since the angle of incidence and reflection are equal. The reflected sound thus passes above the house.
33	Low-frequency sound of truck diffracts around the top of the barriers more than the high-frequency siren, and hence these waves 'bend' towards the house. (Key physics relationship is that the amount of diffraction is proportional to λ/ω)	
34	100 Hz	f = 1/T where T, the period, is 0.01 s
35	B	The air in the room does not move with the wave, but vibrates horizontally, since sound is a longitudinal wave.
36	1.2 m (both)	The distance between wave crests or compressions.
37	Georgio's group has a better model with the slinky, since they are representing a longitudinal wave with the vibrations back and forth in the direction of wave travel. Their model helps explain the bunching up and spreading out of air to create changing pressure in the wave.	
38	C	The wave shifts quarter of a wavelength to the right.
39	• as the sound travels out from a small source in all directions, the energy is spread over an increasing wavefront. • Inverse square law could be quoted	*(diagram: spherical wavefronts from source S)*
40	D	the wave will have moved one-quarter of a wavelength to the right
41	0.60 m	distance between adjacent peaks (or troughs)
42	model B	because it represents a *longitudinal* wave, where the vibration of the particle of the spring are *parallel* to the direction of propagation of the wave (in sound waves the molecules also vibrate in this fashion)

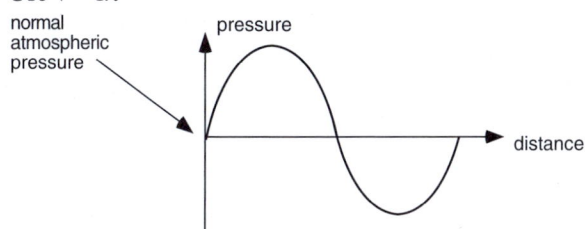

Answers

43	about 20 cm	read from diagram
44	80 cm s^{-1}	use v = fλ
45	200 Hz	Use f = 1/T
46	1.7 m	Use v = fλ
47	A	There must be a time lag for the wave to reach the microphone; this lag = 0.005 s (from speed of sound); the pressure cannot jump straight to a maximum.

Sound intensity and level

48	3.2×10^7	$75 = 10 \log I_1/I_2$, $\Rightarrow I_1/I_2 = 10^{7.5}$
49	30 dB	Reading off the graph, the minimum sound level for 100 Hz is 65 dB, while for 500 Hz it is only 35 dB
50	115 dB	read the difference from the graph
51	3.2×10^{11}	a sound level difference of 115 dB converts using the formula: $\Delta L = 10 \log_{10}(I_1/I_2)$
52	Compared to loud volume sounds, where the curve is relatively flat, at low volumes the ear is less sensitive at low and high frequencies. The loudness button compensates for this.	
53	2.5×10^{-9} W m^{-2}	The intensity drops off with the square of the distance (an inverse square law); $= (1.0 \times 10^{-6}) \times (0.1/2)^2$
54	30 dB	$\Delta L = 10 \log_{10}(I_1/I_2)$
55	3 500 Hz	minimum of the graph
56	600 – 10 000 Hz	read across the graph at the 1.0×10^{-11} W m^{-2} level
57	5 000 Hz	Must be at the minimum of the curve.
58	0 dB	no difference in intensity, hence no difference in levels
59	95.4 dB	Use L = 120 + 10 log I
60	1.4×10^{-4} W m^{-2}	Inverse square law; sound intensity drops by a factor of 5^2.
61	6 dB	Use $\Delta L = 10 \log_{10}(I_1/I_2)$
62	63 dB	$10 \log (2 \times 10^{-6}/1 \times 10^{-12})$
63	9.5 dB	The inverse square law for intensity means the sound intensity has been reduced by a factor of 9, giving a reduction in level of 10 log 9
64	20 dB	Intensity at 500 Hz is 10^{-8} W m^{-2}; at 2500 Hz it is 10^{-10} W m^{-2}, then use $\Delta L = 10 \log_{10}(I_1/I_2)$
65	30 dB	Use L = 10 log (1000 × 10^{-12}/10^{-12})
66	7 dB	10 log 5
67	– 2 dB	read from the graph.
68	600 Hz	The point at which the output graphs cross.
69	4.0	$6 = 10 \log I_1/I_2$
70	600 – 8000 Hz (approx)	read from graph
71	6×10^{-3} W m^{-2}	Inverse square law gives $I_{Jimi}/I_{Tina} = (35/10)^2$, $(I_{Tina} = 5\times10^{-4}$ W m$^{-2})$
72	7.7 W	power crossing the area of a sphere of 35m radius is given by the intensity × area. $P = 5 \times 10^{-4} \times 4\pi(35)^2$, and this gives the total power output.
73	87 dB	10 log (5×10^{-4}/10^{-12})
74	DBAC	
75	11 dB	10 log (6×10^{-3}/5×10^{-4})

172

Answers

76	5 dB	Reading from the graph.
77	3.2×10^{-12} W m^{-2}	10 log I_1/I_2
78	65 ± 5 dB	Reading from the graph: 98 dB – 33 dB.
79	3.2×10^6	65 = 10 log I_L/I_S
80	9.0 dB	10 log 8, since 8 speakers will have 8 times the sound intensity.
81	0.32 W m^{-2}	Power/Area of 5 m sphere; $100/4\pi(5)^2$
82	Louder	L = 10 log $0.32/10^{-12}$ = 115 dB, which is much louder than the 60–70 dB one would hear in a noisy classroom.
83	6 dB	10 log 4 is the increase associated with a $\times$ 4 increase in intensity.
84	8.5 dB	10 log($5 \times 10^{-8}/7 \times 10^{-9}$)
85	4.3×10^{-8} W m^{-2}	The difference between the two intensities: $(50 - 7) \times 10^{-9}$
86	$I_P - I_A - I_B$	This represents the sound energy lost from the system, being the initial – final sound energy.
87	C	The sound energy is transformed to heat energy.
88	10 dB	If 90% absorbed, the intensity is reduced by a factor of 10: ΔL = 10 log 10
89	99%	$\Delta L = -20 \Rightarrow I/I_{incident} = 10^{-2}$. Intensity is reduced by a factor of 100.
90	C	The inverse square law accounts for most of the decrease.
91	90 dB	Use L = 10 log $(1 \times 10^{-3}/1 \times 10^{-12})$
92	1.0×10^{-7} W m^{-2}	Inverse square law; $I_1/I_2 = r_1^2/r_2^2$ (drops by a factor of 10^4).
93	10 dB	From the graph, the 20dB sound at 200 Hz must be reduced to 10dB to be at the threshold of hearing at 600 Hz.
94		Sound intensity in the hall will be louder since the student will hear not only sound directly from the band, but also sound reflected back and forth from the ceiling and walls.
95		The graph shows that the high- and low-frequency sounds need to be of greater intensity to be heard. Therefore, if in the open there is a lower intensity across the frequency spectrum, high- and low-frequency sounds may drop close to or below the student's threshold of hearing.
96	A	B and D are much too soft, but C is too loud (probably over threshold of pain)
97	96 dB	Use L = 10 $\log_{10} I$ + 120
98	2.5×10^{-4} W m^{-2}	Use inverse square law – intensity a factor of 4^2 less than at P.
99		Less if reflected sound interferes *constructively*; more if it interferes *destructively*. However the overall amount of sound energy will increase, so an increase in the level is the best answer.
100		4000 Hz marks the minimum threshold – hence the max. sensitivity of the ear.
101	20 m	The intensity has dropped by a factor of four, which means that the intensity must have dropped by a fator of two (inverse square law)
102	- 6 dB	- indicates a drop; 6 dB comes from $\Delta L = 10 \log_{10}(I_1/I_2)$
103		Sound takes time to travel from the instruments to the listeners. Moreover, some of the sound will have come by relatively long indirect paths via multiple reflections on the hall. The average ear (as shown on the graph) will be less sensitive to high and low frequenies, so even when these are still arriving they may not be heard.
104	~3 500 Hz	This is the frequency that the ear is most sensitive to.

Answers

Standing waves and resonance

105 At the centre point there is a point of constructive interference (antinode) because the sound from each speaker has travelled exactly the same distance. As you move away from this point, the sound from each speaker travels a different distance, resulting in a *path difference*. When the path difference is equal to a half wavelength, destructive interference occurs (a node) resulting in a point of softer intensity.

106
- wavelength of the sound is 1m (from $v = f\lambda$)
- for constructive interference, PD must be = 1m
- moving 0.5 m towards either speaker will accomplish this

107 5.6 m   the distance between two nodes is half a wavelength

108
- the wavelength of a 55 Hz sound is 6.2 m
- this is double the length of the tunnel
- this corresponds to a standing wave with pressure nodes at each end of the tunnel; see diagram below:

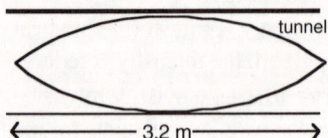

[diagram: standing wave in tunnel, 3.2 m long]

109
- resonance occurs when a standing wave forms inside the resonating system
- standing waves only exist for particular frequencies, determined by the dimensions of the resonating system and the speed of sound
- the tunnel 'selects out' the particular frequency(ies) that will form a standing wave; the other frequencies have little effect.

110
- the strong resonance occurs when the fundamental frequency is excited
- the weaker resonances occur at the third and fifth harmonics
- the second and fourth harmonics do not occur because the bottle behaves like a tube closed at one end and open at the other

111
- filling the bottle partially with water effectively shortens the 'tube'
- the effect of this should be to shorten the wavelengths at which the 'tube' resonates, and hence raise the resonant frequencies

112 17 cm   pipe is closed at one end; $\lambda = 4L$ (then use $v = f\lambda$)

113 B E   pipe is closed at one end; only *odd* harmonics are present

114 B   open end must be a pressure *node*; closed end must be an antinode; pattern must be the simplest (since fundamental)

115 A pipe closed at one end must have a node and an antinode at the ends. The possible wavelengths of standing waves set up in such a tube, of length L, are 4L, 4L/3, 4L/5 ... The first overtone thus has a frequency three times that of the fundamental, 1320 Hz.

116 A flute can be modelled as a pipe open at both ends. Its first overtone will involve a standing wave which has a pressure node at each end and at the centre:

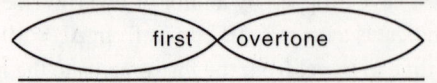

[diagram: first overtone standing wave]

It will have a wavelength L, half that of the fundamental. Its harmonic frequencies are f_0, $2f_0$, $3f_0$ ... In this case 440, 880, 1320 ... Hz.

117 C   A 65 Hz note has wavelength 334/65 = 5.1 m. The length of the column must be half this, at 2.56 m.

118 130, 195 Hz $2f_0$, $3f_0$.

Answers

119		'Resonates' refers to the pipe 'amplifying' more strongly the sound from the loudspeaker. This occurs when a natural frequency of the pipe coincides with the loudspeaker frequency. When this happens a standing wave is likely to be set up in the tube. Now energy can be easily transferred into the tube and a large amplitude vibration is set up. Hence the 'amplifying' effect.
120		• A standing wave is set up in the pipe. • This can be thought of as the superposition of waves travelling in opposite directions due to reflections from the ends. At certain points $\lambda/2$ apart, these interfere destructively, with a maximum from one always coinciding with a minimum from the other. • At these points there is low intensity.
121	5.0 m	Twice the internodal distance.
122	2.5 m	The antinodes are equally spaced, midway between the nodes.
123	C	If there are three nodes, as well as those at the open ends, the length contains four 2.5 m standing wave sections.
124	70 Hz	$f = v/\lambda = 350/5$
125	B, D	These resonant frequencies would correspond to standing wave patterns as shown in the diagrams following.

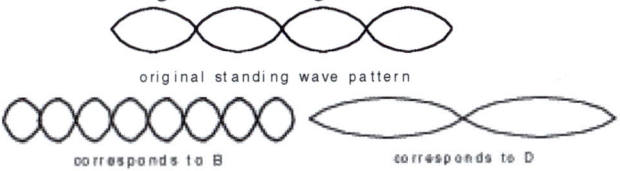

original standing wave pattern

corresponds to B   corresponds to D

126	C	For a pipe closed at one end the resonant frequencies are 8.8, 26.3, 43.8, 61.3, 78.8 Hz (five resonances); for an open pipe they are 17.5, 35, 52.3, 70, 87.5 (also five).
127	Closed at one end	The fundamental from this has a wavelength equal to 4 L. The standing wave pattern (pressure amplitude pattern) looks like:

A tube open at both ends would have a pattern like:

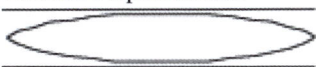

Its wavelength would be equal to only 2L; a 'higher' note (larger frequency)

128	AC	C is correct; combine this with $c = f\lambda$ to give A also.
129	BD	Establish D first, then combine with $c = f\lambda$
130	0.30m	The sequence of wavelengths correspond to the first three harmonics of a closed pipe, with wavelengths 4L, 4L/3, 4L/5.
131		Though a player's lips vibrate at a range of frequencies, the air column of the bugle will 'select' frequencies that correspond to standing waves and amplify these. (A player can adjust their lip frequency to favour resonance at a particular harmonic, which becomes the fundamental of the note produced.)
132		The diagrams should show the possible standing wave patterns (as displacement amplitude diagrams), with nodes at either end of the string:

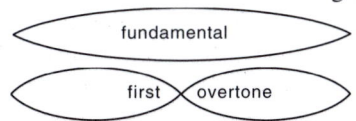

fundamental

first   overtone

These correspond to wavelengths 2l, l, 2l/3 .. so that the frequencies are in the ratios 1:2:3

Answers

133   0.230 m   Resonance will occur each time the tube length increases by $\lambda/2$, so $\lambda = 2 \times 0.115$ m.

134   The air column in the trombone will resonate at particular frequencies (given by $l = n\lambda/2$); all other frequencies from the player's lips will be suppressed; the feedback between the lips–trombone system will cause sound at the resonant frequencies to be selectively amplified.

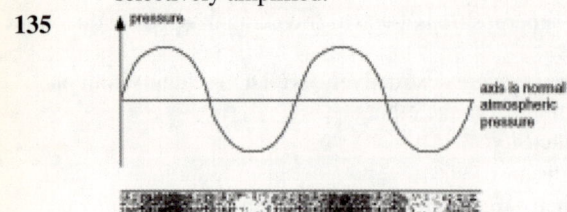

135

136   2.76 m   There are two complete wavelengths within the pipe.

137 4.1×10^2 Hz $v = f\lambda$, largest wavelength corresponds to the lowest frequency.

138   21 cm   For a pipe closed at one end, the resonant frequencies are in the ratio 1:3:5:7 and the longest wavelength is four times the length of the tube.

139   9.3 cm   The next wavelength in the series would be one-ninth the wavelength of the fundamental (0.84/9). (Tubes closed at one end only have odd-numbered harmonics present.)

140   the same   The wavelengths at which resonance occurs depend only on the length of the tube (4L, 4L/3, 4L/5 etc.).

141   There are several possible routes for vibrations to reach the ears of the player: the wall of the flute will vibrate at the frequency, and set up travelling waves in the air, which travel to the ears of the player. The air at the open ends of the pipe will also set up travelling waves outside the pipe (reflection is not perfect at the open ends), and sound may also reach the player's ears through vibrations transmitted through the skeleton of the player (initiated by contact between the player's body and the instrument)

142   2.50 m   One full wavelength is *twice* the distance between two nodes.

143   136 Hz   Use $v = f\lambda$.

144   D   P is between a node and an antinode; hence pressure amplitude is not maximum, but not zero either; also in phase with variation at Q.

145   C   number of nodes in the pattern increases from 2 to 3

146   394 Hz   Use $v = f\lambda$.

147   The wavelength is exactly the same, as resonant wavelengths are determined by the length of the tube (the speed of sound will affect the *frequency*).

148   One open end only   The nodal/antinodal pattern looks like:

N an   N an   N an   N an   N an

- the open end *must* be a node
- the closed en *must* be an antinode

149   1.13 m   The pattern shown above corresponds to 2.25 whole wavelengths, hence $2.25\lambda = L$; since $\lambda = 0.5$ m (from $v = f\lambda$) this gives the answer.

176

Diffraction and interference

150	B	she is walking along an *antinodal* line, and simply getting closer to the source (hence louder)
151	2.7 m	• the distance from S_2 to D is 16 m (from the diagram) • the distance from S_1 to D is 20 m (using Pythagoras) • the *path difference* between these two paths is 4 m • this must be equal to $1.5\lambda = 4.0$ m
152	0.33 m	• the speed of sound is the same for both frequencies • hence $f_1\lambda_1 = f_2\lambda_2$

153   • key principle here is *diffraction*
• the amount of diffraction depends on the wavelength and the aperture; proportional to λ/w.
• in this case, there is less diffraction of the short wavelengths (high frequencies) from the larger aperture speaker; for the large speaker $\lambda/w = 0.33/0.35 = 0.94$; for the smaller speaker $\lambda/w = 6.6$

154   Fred is walking through the maxima and minima of a standard two source interference pattern. The maxima in this two source pattern are due to sound from one loudspeaker travelling a different distance from the sound from the other speaker. When this path difference is equal to 0, λ, 2λ, 3λ, ... the result is constructive interference, or a loud spot. When the path difference is $\lambda/2$, $3\lambda/2$, $5\lambda/2$, ... the interference is destructive, leading to a soft spot.

155	16.5 m	The *path difference* must be one wavelength. The wavelength of the sound is 0.5 m (from $v = f\lambda$); hence the distance involved is 16.5 m.

156   The intensity drops as they move through the door because they are no longer receiving sound reflected from the walls. The door acts like a single slit through which the sound is diffracted, so the intensity drops of as they move towards Y. The intensity will decrease at a greater rate for high frequency, short wavelength sound, since the diffraction pattern is narrower.

157	A	The ratio of λ/w for 400 Hz sound wave diffracted by the door is 1. This means 'complete' diffraction through 180°. At A and C, listeners will be within the central diffraction maximum only for f< 400 Hz.

158   If the speakers are out of phase, the two wave trains arriving at X will have travelled the same distance and will still be out of phase. A compression from one will coincide with a rarefaction from the other. The two waves will superimpose to give zero pressure variation.

159	Minimum	The two speakers are at 1.6m and 2.0m from Junia. Since the difference in distance is 0.4m, or half a wavelength ($\lambda = 320/400$), the waves will arrive out of phase.
160	A	For 800 Hz, $\lambda = 0.4$ m. Extra distance to bottom speaker is 0.4 m, equal to one whole wavelength.

161   The width of the diffraction pattern from the speaker is increased if the opening is narrowed, since the spread is proportional to λ/d.

162	Approx. 5 cm	For a good spread, λ/d should be close to 1, implying $d = \lambda = 325/6500$ m.

163   As with previous questions, the reason has to do with the spread of the diffraction pattern. A low-frequency audible wavelength will have a wavelength (e.g. for a 40 Hz sound $\lambda = 8$ m) much larger than the aperture dimensions and will be strongly diffracted. A high-frequency sound will have a wavelength less than the speaker aperture and will have a narrow diffraction band. For a 16 kHz wave, $\lambda = 2$ cm and the first diffraction minimum for a 5 cm diameter 'tweeter' will be approximately $\theta = \sin^{-1}\lambda/d = \sin^{-1} 2/5 = 20°$.

Answers

164 The sound that passes over the barrier will diffract around it since sound is a wave motion. The amount of diffraction will depend on the width of the barrier and the wavelength of the sound. In general, the amount of diffraction is proportional to λ/w. Low-frequency waves have a much longer wavelength than high-frequency waves, therefore they will diffract more. Hence the barriers work better for high frequencies than for low frequencies.

165
- at the point Z the student is equidistant from each of the speakers
- sound waves are arriving *in phase* at this point
- this is a maximum of sound intensity
- either side of Z the distances from the student to the speakers are unequal
- this means that sound waves have travelled different distances before they reach the ears of the student
- this allows the possibility of *interference*
- when the path difference to the speakers is equal to $\lambda, 2\lambda, 3\lambda, \ldots$ there is *constructive* interference and a maximum is heard
- when the path difference is $\lambda/2, 3\lambda/2, 5\lambda/2, \ldots$ there is destructive interference and a minimum is heard
- in between these points the intensity is intermediate
- hence, as the student walks along he/she encounters a series of maxima and minima in sound intensity

166 20.2 m   Path difference at this point must be $\lambda\ (= 0.2\text{ m})$

167 0.8 m $v = f\lambda$

168 As the frequency is increased and the wavelength shortened, the waves arriving from A and B become progressively more out of phase until at some value of f they interfere destructively at L.

169 1275 Hz   This occurs when the path difference, 0.4m, corresponds to $3\lambda/2$ so the waves are again out of phase.

170 The intensity will go through a series of maxima and minima. It will record three maxima before arriving opposite A, at the second node.

171 3.4 cm, 3.4 m $\lambda = v/f = 340/10\,000;\ 340/100$.

172
- Each part of the speaker can be regarded as a sound wave source, and its limited size gives rise to diffraction effects;
- Diffraction will spread the sound waves by an amount proportional to d/λ;
- For the low frequency, λ is much greater than the speaker diameter and the diffraction maximum will be spread widely, from A to B;
- For $\lambda = 3.4$ cm (wavelength of 10 kHz sound), less than the probable speaker diameter, the diffraction will be smaller and the sound much more directional.

173 1.5 m
- As the wavelength is increased, the interference pattern from the two speakers spreads out;
- At a certain maximum wavelength (minimum frequency) the point M will coincide with the first nodal line;
- At this point $S_2M - S_1M = \lambda/2 = 0.5$m and the waves from S_1 and S_2 are out of phase.
- Use Pythagoras' theorem with $S_2M = 2.5$m.

174 No more intensity maxima will be detected since the 0.75 m shift will place the microphone at the central maximum.

Answers

175
- At chair X the sound is always a maximum since the waves from each of S₁ and S₂ arrive in phase.
- A pressure maximum from S₁ coincides with a maximum from S₂, creating doubly high pressure. The same is true for minima.
- The pressure amplitude is double what it would be for either speaker alone.

176 X: 1.6 m, Y: 1.2 m   The path difference for X is $S_2X - S_1X$.

177 0.8 m   Moving from an antinode to the next node, the path difference changes by 0.5λ. Hence the difference $1.6 - 1.2 = 0.4$ m must represent half a wavelength. Y must be on the second nodal line.

178 If the frequency is doubled, the wavelength is halved to 0.4 m. This means that the path difference for Y, 1.2 m, is now a whole number of wavelengths (3λ) and Y lies on the third antinodal line.

179
- The effect is due to diffraction.
- The extent of 'bending' of sound round an obstacle of diameter d is dependent on the ratio λ/d.
- Dan's head will cause a 'sound shadow' for high frequencies, where λ is approximately the same, or smaller than d.
- For a wavelength of 5 m, which is much larger than d, the wave will diffract round his head so that there is practically no shadow effect.

180 500 Hz   Use $f = 1/T$; $T = 2$ ms.

181 The wave should be identical in shape and shifted 0.5 ms to the right compared to the right ear pressure variation. It should also probably be of a smaller amplitude.

182
- at a minimum of the interference pattern, waves arrive exactly out of phase
- this is true at all times, hence the pressure amplitude remains at zero.

183 2.5 m   path difference $= S_2P - S_1P$

184 1.0 m
- path difference at Q is 2.0 m ($= n\lambda$)
- path difference at P is 2.5 m ($= n - \frac{1}{2}\lambda$)

difference between these must be one half of a wavelength

185 9.1 m $XQ - YQ = \lambda/2$ at the first minimum

186/7 D   high frequencies diffract less than low frequencies (diffraction $\propto \lambda/w$)

188 B   the sound will arrive one-quarter of a period later.

189 at higher frequencies there will be less diffraction (since diffraction is proportional to λ/w), so the sound will not 'bend' around the head as well.

190
- the sounds from the double bass have long wavelengths and low frequencies compared to the short wavelengths and high frequencies of the sounds from the violin
- long wavelengths diffract more readily than short wavelengths
- this is confirmed by the ratio λ/w

191 212 Hz   2.4 m (path difference) must be equal to either $\lambda/2$ or $3\lambda/2$ etc. Only one of these gives a frequency in the correct range (using $v = f\lambda$)

192
- for high frequencies and short wavelengths, the effect of diffraction is certainly considerable, as the ratio λ/w will be small – i.e. much less than 1. Take the example of $f = 6800$ Hz (where $\lambda = 5$___). This is much less than the diameter of the head.
- however at low frequencies diffraction will be considerable; it will not be possible to use this to distinguish the direction. Hence at these frequencies John's explanation is likely to be the better.

Answers

Magnetic and electric basics

193	8.1×10^5 J	Energy = power × time (in seconds)
194	1.0 (Ω)	V = P/I = 15 V
195	5.4×10^4 C	Q = current × time
196	12 000 J	Use area under the graph
197	125 A	Use P = VI, remembering that the battery only draws 1500 W (not 1800)
198	12 J	This is the definition of the volt (one joule per coulomb)
199	12 kJ	Use E = area under graph (note non–zero axis)
200	50 A	Use P = VI
201	750 C	Use Q = It
202	17 Ω	$1/R_{tot} = \Sigma(1/R)$
203	840 W	$P = V^2/R_{tot}$
204	6.0×10^5 C	Q = I × t, where I = 7 A
205	7.3×10^7 J	Power (840) × time (in seconds)
206	0.57 A	I = V/ΣR = 120/210. The assumption that the resistance remains the same for all voltages for these appliances is not really justified.
207	No	The appliances would each have much less than the 120 V for which they were designed to operate: Heater: 69 V, Food processor: 34 V, Microwave: 17 V.
208	20 kΩ	12 kΩ in series with (12 kΩ parallel to 24 kΩ = 8 kΩ)
209	Rise	In the arrangement shown, $V_{CB} = 0.4\ V_{AB}$. With the 24 kΩ resistor removed, $V_{CB} = 0.5\ V_{AB}$
210	B	The magnetic field at P points to the left. The right-hand rule gives the force to be out of the page.
211	A	Use the right-hand rule (or left hand rule for electrons)
212	D	Use the right-hand rule. It can be used assuming positive charges are flowing up the conductor.
213	around 2 N	Use F = BIL. Reasonable estimates would be B = 20 μT, L = 10 m.
214	C	If the conductor is bent towards the north, it will now not be perpendicular to the magnetic field, and the force will be less than before.
215	10 A	I = F/LB = 3/(2 × 0.15)
216	to the left	The force on coil 1 is directed away from coil 2. The current directions are opposite in the two coils, and they act like two magnets with opposing (south) poles.
217	AC	When the wire is pushed sideways, it must be due to current flow. The moving electrons in the wire are acted on by a magnetic force. The formula BIL is a special case where the wire is perpendicular to the field.
218	D	The magnetic fields add as vectors. The fields due to the two magnets will cancel at the point P.
219	38 A	Use $P = I^2R$
220	420 V	Use $P = V^2/R$
221	54 A	$I_p = I_{RMS} \times \sqrt{2}$
222	593 V	$V_p = V_{RMS} \times \sqrt{2}$
223	36 W	P = VI
224	2.6×10^5 J	E = P × t (t in seconds)

… Answers

225	2000 J	$V_{rms} \times I_{rms}$
226	0.5 A	Using I = P/V
227	320 Ω	R must be twice the size of the resistance of the lighting system (voltage divider). Since resistance of the lighting system is 160 Ω (from V = IR); R = 320 Ω.
228	120 W	P = VI = 240 × 0.5
229		The resistor uses up energy as heat. With a step-down transformer, only 40 W need be drawn and *all* the energy can be dissipated in the lighting system.
230		

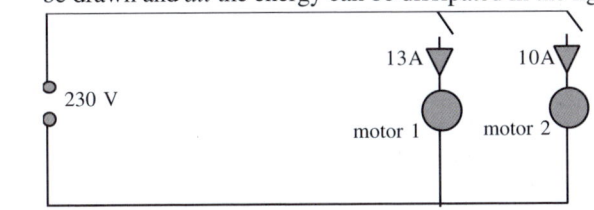

[circuit diagram: 230 V supply connected to motor 1 (13A fuse) and motor 2 (10A fuse) in parallel]

231	5.29 kW	P = VI = 230 × (10 + 13)
232	230 W	Either calculate the power used by the motors 220 × (13+10) = 5.06W and subtract, OR Calculate the power directly: there is a 10V drop along the wire and a current of 23A.
233	C	C is the closest alternative; the field will be strong inside the solenoid and weak outside; the 'right hand grip' rule gives the correct direction.
234		The correct combinations are: AF; BD; CE.
235	2.5 C	Use P = QV/t
236	A	Use right-hand grip rule.
237	30 ohm	Use P = VI to obtain RMS current (4A); total resistance = 60Ω, hence resistance of one element is 30 Ω.
238	4	Since $P = V^2/R$; the ratio will be the same as $R_{series}/R_{parallel}$; this is 60/15 = 4.
239		

[circuit diagram showing a battery, switch, resistor and light bulb in series]

240	0.001 Ω	Use R = V/I
241	6.25×10^{-17} s^{-1}	Current = Q/t, where Q = ne (n is the number of electrons; e is the charge on one electron.) Take t = 1 s.

Motors

242	B	Use the right hand rule on the current element near the arrow or treat the coil as a small magnet with field into the page, which swings to line up with the field.
243	BDF	The torque on the coil is proportional to I, n and the area of the coil.
244	D	The electromagnets should act exactly like the permanent magnets.
245	C	B field is to the left; this gives a force out of the page (use RHS rule).
246	No	Reversing the current flow in one of the magnets would reverse the field so that the two fields oppose and cancel at the coil position. There might be some small field left at the edges.

Answers

247 The connection of AC to just the armature coil would cause the motor to 'just vibrate', as the magnetic force would change direction as the AC changed direction. However, as the field coils are also connected to the same AC supply, the armature current and the field current both change direction together, at the same time. This causes *no* change in the direction of the magnetic force. As far as the magnetic force is concerned, the motor might just as well be being supplied with DC.

248
- the 'shoestring motor' should *not* work because there is no mechanism for reversing the current – no commutator
- the magnet should have side A as a pole and the coil should rotate to a horizontal plane and stop there.

249 D — Successive pulses of current at the right times (and not at others) could kick the motor round. This is possible when the motor 'bounces' off the contact points, making contact (on the average) every *half* turn. This avoids the need for a commutator.

250 A — The force on XY will be in the opposite direction to Q and on the top side will be in the same direction as Q.

251 0.12 N — nBIL = 10 × 4 × .02 × 0.15

252 D — The forces on each side will be inward, tending to collapse the coil rather than turning it from this position. There will be no net torque on the coil.

253 The force on WX will be vertically upwards; that on ZY will be vertically downwards.

254 0.47 N — Use F = nBIl (watch the units of B, they are in mT not T. They must be converted before calculation).

255 0 N — There is no force when the current and the magnetic field are parallel as in this case.

256
- current will flow from **b** to **a** along the side **ab**
- the force on side **ab** will be *into* the page (RHS rule)
- the force on **cd** will be *out* of the page
- there will be net torque on the coil, which should initiate rotation

257
- the torque in the previous question will rotate the coil through 90^0
- at this stage the forces on the sides **ab** and **cd** will be 'in-line' and there will be no torque on the coil
- what is needed at this stage is a device or arrangement to reverse the flow of current in the coil
- if this occurs, a momentum carrying the coil past the '90°' position will ensure that the torque is in the same direction as the starting torque
- the device that causes this reversal of the current at this part of the coils rotation is called a *commutator*

258 0.08 N — nBIL = 20 × 2.0 × 0.02 × 0.1

259 zero — The horizontal sides are parallel to the field and there is no force.

260 B — RH rule, to give a force on AD into the page.

261 0.060 N — Use F = nBIl

262 0 N — The wire is parallel to the magnetic field on this side.

263
- when the coil is horizontal torque will be unchanged
- when the coil is *not* horizontal, torque will be greater, because the magnetic field is at right angles to the wire,
- whereas with the flat pole pieces the magnetic field was *not* at right angles, thus reducing the force in this position, and hence the overall average torque

Answers

264	From K to L	The BIl force must be up to make the coil turn as shown (use RHS rule)
265	4.0A	Use $F = BIl = 0.60$
266	a.	The coil will flip in the direction shown since:

- the force on side KL is up;
- the force on side NM is down;
- therefore there is a net torque on the coil.

      b.   The coil will remain stationary, since:
- the forces on KL and MN are up and down, depending on the current direction, causing a squeezing or expanding effect but not a rotation;
- the same argument applies to KN and LM.

267	B, B, B	In each case the right hand rule gives the BIl force to be vertically down in the diagram, causing the loop to rotate.
268	E, E, NF	In the first two orientations the right hand rule gives the BIl force to be out of the page, toward the centre of the loop. In orientation 3 the wire is aligned with the field and there is hence no force
269	9×10^{-4} N	use $F = BIl$
270	A, A, A	force is always at right angles to field and current; RHS rule
271	NF and C	in (a) current is parallel to field, use RHS rule for (c)
272	0.15 N	Use $F = BIl$
273	0 N	KL is parallel to the magnetic field, hence no force
274		Force on JK is down, on LM is up, hence no torque, but forces tend to compress coil. On KL force is out of the page, on JM inwards. Again, force is a compressing one, no torque again.
275	B	Use the right-hand grip rule on the field windings
276	0.9 N	Use $F = nBIl$
277	A	Use the RHS rule on the sides JK and ML
278–281	G, I, H, G	Use RHS rule (or Fleming's right-hand rule)
282	B	(this is true provided there is no friction of any kind; if there is a lot of friction, A would be a better answer). The reasons for B are:

      • initially there is a torque which continues until the coil has turned through 90 degrees
      • at this stage the torque drops to zero (and then reverses); however, the coil by now has gained momentum, and the reversed torque merely slows and then stops the coil after a half turn
      • the (reversed) torque now accelerates the coil back in the direction it has come from
      • the process continues...

283	D	He has now built a 'proper' DC motor; the commutator reverses the direction of the current at the right time so as to ensure continuous motion in the same sense.

Generation principles

284	Sinusoidal graph with amplitude 1.5 V and period 2.5s.
285	The emf is equal to the negative time rate of change of flux. The flux will follow a sinusoidal curve. The emf will be positive or negative depending on whether the flux is decreasing or increasing. It will be a sinusoidal variation 90° out of phase with the flux variation.

Answers

286	D	The flux will change direction from the 'coil's point of view', depending on which way the field lines cut through the coil.
287–288	B	The flux will change more rapidly as the magnet falls faster – hence B rather than A. It will not change sign because the field is always in the same direction.
289–290	C	The emf is equal to the negative time rate of change of flux. The flux is increasing during the first part; decreasing during the second part as the magnet falls away. Hence there must be a change in the direction of the induced emf (voltage).
291	6.7 V	$\varepsilon = n \, \Delta\phi / \Delta t = 100 \times 0.05 \times 0.2 / 0.15$
292		The emf is proportional to the negative of the gradient of the graph of the flux against time (from Faraday's Law: emf $= - n \, \Delta\Phi/\Delta t$). Between t_1 and t_2 the gradient is negative and constant (hence the emf); at other times the gradient is zero and so is the emf.

293 The graph was at first: The new graph will look like:

The change occurs simply because the gradient of the flux graph will be one-quarter of what it was before.

294	1.2 V	Use Faraday's Law: emf $= - n \, \Delta\Phi/\Delta t$

295 The emf induced between t_1 and t_3 is constant because of the constant slope of B. Whether B is positive or negative does not matter.

296 A motor consists of a coil which rotates between magnets when a current is supplied. If it is spun without a power source it will act as a generator since the flux through the coil will be changing. This creates an induced emf which will light the globe.

297
• if the switch is closed current will flow
• the coil, because of this current, will experience a BIL force which opposes its motion by creating a torque.

298	C	The weight will accelerate, unless friction is too great. A faster fall will spin the coil faster and produce a larger emf.
299	1.4 W	$P = I^2 R$

300
• the induced voltage is given by emf $= - \Delta\phi/\delta t$
• this is the rate of change of flux
• this is the gradient of the flux–time graph
• this gradient is constant, hence the emf is also

301	A	$\Delta\phi$ is twice as large, Δt is half the value, hence emf is $4V_0$
302	5×10^{-4} m^2	use emf $= \Delta(BA)/\Delta t$
303	B	The induced voltage is proportional to the rate of change of flux through the coil. The flux is proportional to the field. Hence the field changing with constant slope gives a constant induced voltage. $V_{ind} = - d\phi/dt$

Answers

304 4.7×10^{-5} Wb $\phi = B \times$ area of loop $= 0.15 \times \pi r^2$.
305 7.2×10^{-5} Wb $\phi = B \times$ area of loop
306

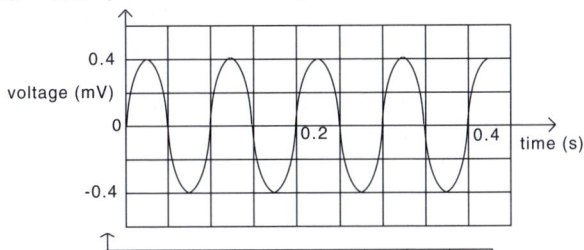

307

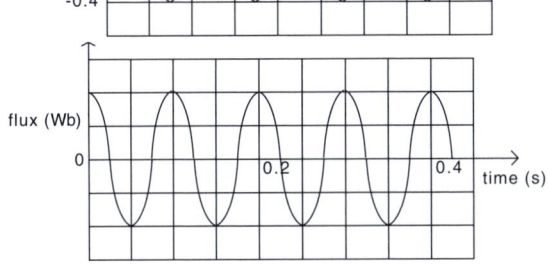

308 2×10^{-5} Wb $\phi = B \times$ area of loop
309 5×10^{-4} V $E = \Delta\phi/\Delta t$
310 B   negative gradient of flux graph gives emf
311 A   The flux through the coil changes, causing an emf (Faraday's law).
312 Something close to the diagram on the right.
The graph is the negative slope of the flux graph. The emf will be large at t = 0 and cross the axis at 0.2 s when the flux peaks (points of zero gradient).

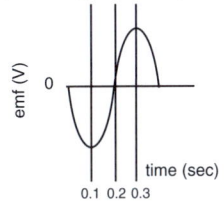

313 3.5 V $\varepsilon = \Delta\phi/\Delta t = 50 \times 0.014 / 0.2$
314 0.78 A $I = \varepsilon/R$
315 D   The emf is equal to the *rate of change* of the magnetic flux (or field in this case); or the *gradient* of the field-time graph.

Transformers, transmission and consumption

316 558 $N_p/N_s = V_p/V_s$
317 E $N_p/N_s = V_p/V_s$
318 0.83 $P = VI$
319 There will be an inevitable power loss due to stray currents in the core and heating. In this case the transformer has 83% power efficiency.
320 20 $N_p/N_s = V_p/V_s$
321
- the secondary voltage is due to electromagnetic induction
- this is caused by a changing flux through the core of the transformer
- the emf from Faraday's Law: $\varepsilon = n \Delta\phi / \Delta t$
- the changing flux is created by the current in the primary coil
- if this does not vary, there will be no induced emf.
- the output will be zero.

322 22   Use $n_1/V_1 = n_2/V_2$
323 123 A   Assume input VI = output VI

Answers

324 • there is a 5 000 V drop in the voltage between the output of the switchyard transformer and the next transformer
• this is caused by resistive losses in the transmission lines joining these transformers
• the relevant formula is V = IR, where V = 5 000 V, R is the resistance of the transmission lines, and I is the current flowing in the lines

325 • normally all the current that flows into an appliance should flow out of it
• if the currents are unequal, then some of the current is finding another route out of the appliance
• this route might be through a human, thereby endangering that person
• in any case, an inequality indicates a fault in the appliance, such as a failure of insulation

326 • the active current will cause flux in one direction through the core
• the neutral will cause an exactly equal amount of flux through the core
• since the active and neutral wires are wound in opposite directions around the core, these fluxes will be in opposite directions and exactly cancel

327 C   • the inequality ensures a net flux through the core
• the AC nature of the current ensures that this will change as the AC changes with time

328 • the voltage at the house will be less than 240 V by an amount V = I
• I is the current drawn by the house; R the resistance of the supply cables
• as the current varies, so will the voltage at the house

329 $0.33\ \Omega$ V_{loss} in wires = 15 V; use $V_{loss} = IR_{wires}$

330 10.5×10^9 J   92 days $\times$ 32 kW h $\times$ 3.6 (conversion of kW h to MJ)

331 There is a voltage drop due to resistance in the wire, $\Delta V = RI$.

332 AD   Current is not lost, but energy is transformed to other forms (thermal energy).

333 30.4 kW   Power loss = ΔV (0.2 kV) $\times$ I (10 MW/66 kV = 152 A)

334 $1.3\ \Omega$ R = V/I

335 0.44 MW   R I^2, where R = 80 $\times$ 0.24, I = 152 A.

336 6 cents   60 $\times$ .0085 $\times$ 11.7

337 50 A   I = P/V = 10 000/200

338 $0.6\ \Omega$ R = voltage drop (30 V) $\div$ current.

339 1500 W   P = $\Delta V \times$ I

340

341 2.5 A   I = 10 000/4000. One twentieth the previous value.

342 199.9 V ΔV along wire = 2.5 $\times$ 0.6 = 1.5 V. When stepped down this becomes 3998.5 / 20.

343 3.8 W RI^2

344 0.0025   3.8/1500

345 24 turns   from $V_1/V_2 = N_1/N_2$

346 C   Current must be AC; average value of AC = 0; peak value = 3.5 $\times \sqrt{2}$ A.

347 $0.29\ \Omega$ The voltage drop in the wires is 1.0 V; then use V = IR

348 3.5 W   Use P = I^2R or P = ΔVI

349 E   Use P = V^2/R; must be *at least* 110 V

186

Answers

350	20 A	Thomasina draws 20 kW (200 A × 100 V) from the transformer, implying I = P/V = 2×10^4 / 1000.
351	19.5 × 10^3 kW h	Area under curve from 1 pm to 4 pm = 6.5 × 10^3 kW × 3 h.
352	0.025	$n_s/n_p = V_s/V_p$
353	16W	Subtract I^{2}R, power dissipated in the resistors, from power VI from transformer, OR calculate the voltage drop across the light and use VI.
354	D	The current must alternate, with peak current √2 times RMS current.
355	6.0 kW	P = VI= 240 × 25, since current supplied = 9 + 12 + 4 A
356	1.0 kW	Power supplied at the tow = 200 × 25W = 5.0 kW. The difference is lost in the cables.
357	1.6 Ω	Power lost P = RI^2.
358	660 turns	$N_p/N_s = V_p/V_s$ giving N_s = 1440 x 110/240
359	0.92A	The power drawn will be that supplied to the globe, 220W. Use P = VI with V = 240.
360	480W, 220W	First method: P = VI with I = 240/(55 + 65), or I = 110/55 (ie. I = 2A) [Second method: see above]
361	9.94 × 10^9 J	92 days × 30 kW h × 3.6 (conversion from kW h to MJ)
362	$2.30	Energy consumed = 2.4 kW × 8 hours @ $0.12
363	5 hours	Extra 20 kW h is explained by a 4 kW appliance operating for 5 hr.
364	• The transformers allow the power to the pump to be delivered at a higher voltage; hence less current is flowing along the wires, and hence less power I^{2}R is lost as heat in the wires. • T$_A$ is a step up transformer. T$_B$ is a step down transformer.	
365	6000V	Power to T$_B$ equals power to the pump. V × 0.8 = 240 × 20
366	2500 turns	$N_p/N_s = V_p/V_s$ giving N_p = 100 × 6000/240
367	2.6W	P = I^{2}R
368	400W	P = I^{2}R
369	A	kW h is a unit of *energy* (from E = P × t)
370	8 h	kW h = no. of kW × no of hours
371	300 W	use P = I^{2}R
372	• to reduce the power loss need to reduce either I or R (P = I^{2}R) • if T$_1$ is a step-up transformer, it increases the voltage but decreases the current, without affecting the power transmitted • this implies that T$_2$ must be a step-down transformer to reduce the voltage to that suitable for domestic appliances	
373	0.25 A	use P = VI
374	909 turns	use $n_1/n_2 = V_1/V_2$
375	6 W	use P = VI
376	7500 A	since transformers are ideal, P$_{in}$ = P$_{out}$; hence $V_1I_1 = V_2I_2$
377	7.2 × 10^5 W	use P = I^{2}R
378	498 kV	V$_{loss}$ = IR = 2400; hence V = 500 000 – 2400 = 498 kV
379	200 A	$I_s/I_p = V_p/V_s$ (provided the transformer is 100% efficient)
380		Since the secondary voltage is fixed at 12 V, a high resistance will inhibit the current value (I = V/R).
381	20:1 *or* 20	$N_p/N_s = V_p/V_s$

Answers

382	8 W	Use $P = V^2/R$
383	11 V	• total resistance of wires = $32 \times 0.05 = 1.6\ \Omega$
		• total resistance of circuit is now $19.6\ \Omega$
		• lights will now get a fraction of the full 12V; = $12 \times 18/19.6$
384		• light 2 receives the full 12 V output of the transformer
		• light 1 receives *less* than the full 12 V, because some voltage is dropped over the 16 m long wires (see answer to previous question)
385	D	Provided we ignore internal resistance of the transformer secondary, the voltage across light 2 will not change; as its resistance will be the same, so will the power delivered to it (i.e. 8 W).
386	0.5 A	Use $P = IV$
387		Some energy was being 'lost' in the resistance of the long wires between the two transformers.
388	200 V	$V_{out} = V_{in} \times 240/12$
389	0.42 A	$I_{out} = I_{in} \times 12/240$
390	$0.24\ \Omega$	Use $R = \Delta V/I$
391	B,C	Lowering resistance would clearly lower the losses. However, using the different transformer ratios would ensure the transmission lines ran at 24 V (rather than 12 V). This would result in smaller currents and lower losses (from $\Delta P = I^2 R$)

Electronics basics

392	139 Hz	$f = 1/T$
393	0.85 V	$V_{rms} = V_{peak}/\sqrt{2}$
394	1.2 V	Read from diagram.
395	C	B is possible but gives wrong DC value.
396	12 V	Use voltage divider formula *OR* simply divide the 12 V up in proportion to the sizes of the two resistors. Since they are the same the 12 V must divide equally.
397	8 V	As above
398	16 V	As above
399	16 V	As above
400	0 V	The direct connection to earth ties the voltage to zero.
401	$20\ k\Omega$	Use voltage divider formula.
402	$32.5\ k\Omega$	Use voltage divider formula.
403	0.83 V	Use voltage divider formula.
404	7.1 V	$V_P = V_{RMS} \times \sqrt{2}$
405	4 ms	Use $T = 1/f$
406		

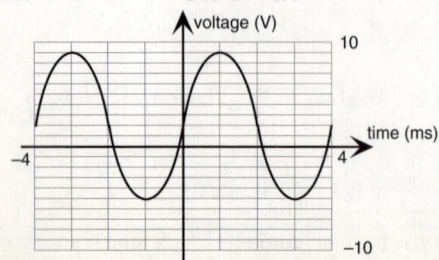

407	$600\ \Omega$	read from graph.

Answers

408	25°C	The resistance is $3.0/5 \times 10^{-3} + 600\ \Omega$. The temperature is read from the graph.
409	decrease	• as temperature increases, the resistance of the thermistor will drop • hence the voltage at X will drop (voltage divider formula)
410	4.8 V	Use voltage divider formula.
411	10 V, 10 μA	Read from graph at intercepts.
412	42 μW	Use P = VI
413		When the voltage is a maximum, the current is zero (hence the power output is zero). When the current is a maximum, the voltage is zero (as is the power output) Hence non-zero values will only occur when the current is less than the maximum and the voltage is also less than the maximum.
414	5.0 V	Use voltage divider formula.
415	V_{out} increases	• as light level increases, R of LDR reduces • hence LDR has a smaller proportion of the voltage across it (see voltage divider formula) • hence V_{out} increases
416	0.8 lux	• resistance of LDR must be also 1000 Ω • read from graph
417	200 mA	The total resistance is 60Ω in this circuit.
418	4 V	The combined resistance of the two resistors in parallel is 10 Ω, making a total resistance of 30 Ω. The current divides so that 200 mA goes through R_2.
419	0.6 s	3 squares; each square is 0.2 s.
420	100 beats/min	Divide 60 by 0.6.
421	42 V	Convert to peak ($\times \sqrt{2}$) then double
422	50 mA	V = IR and RMS values; R = 300 Ω (3 × 100 Ω in series)
423	6.0 V	Total R is now 250 Ω; I = 60 mA; hence V= IR gives 6.0 V
424		Current at I = 0.8 V must be 80 mA (from V = IR); each square is 20 mA
425	D	At V = 0.8 V, I = 80 mA, but at 0.4 V, I is < 5 mA; hence D
426	45 Ω	L_2 has 9V across its terminals. The current through the circuit can be found by considering V = RI for L_1; 3/15 = 0.2A
427	3.5 V	$V_{rms} = V_{p-p} / 2\sqrt{2}$
428	50 Hz	T = 1/f
429		Use voltage divider formula.
430	10 Ω	R_{total} = 3 R; using V = IR implies 9 = 0.3 (3R)
431	C	In upper circuit I = 0.3, V = 3, hence P = VI = 0.9 W In lower circuit V = 3, R = 10, hence P = V^2/R = 0.9 W
432	0.10 V	This is a voltage divider with 10 ohm and 40 ohm. Use the voltage divider formula.
433	C	The current must be the same (two resistors in series)
434	63 mW	The power is shared equally, since both resistors have the same value. The total power is given by P = V^2/r = 0.25/20 = 125 mW
435	70 degrees	R = V/I = 400 Ω, hence read from graph.
436	10.5 V	must be battery voltage (12) minus 1.5 (LED voltage)
437	0.03 A	$I_{LED} = I_R$; I_R = V/350

189

Answers

438	B	• in A, open belt causes open cct, therefore no light • in B, open belt causes a normal LED circuit, hence LED lights • in C, the LED is reverse biased, so it never works
439	0 W	P = VI; no current, no power.
440	1.7 mA	Read from graph.
441	2.8 V	Read from graph.
442	4.2 mW	P = VI
443	1.9 kΩ	R = V/I
444	4 ±.2 V	Trial and error using P = VI from graph.
445	10 ms	One period of 50 Hz is 20 ms (use f = 1/T); hence half a period will be 10 ms. The difference between t_0 and t_1 on the screen is half a period.
446	339 V	• one division is the peak value (definition) • peak value of 240 V RMS is 339 V
447		The graph indicates that the resistance of the thermistor depends on temperature; this means that when a current flows through it, the resistance is likely to change as the thermistor heats due to the current. Hence the thermistor must be *non-ohmic*. (Note that the effect might not be very significant, but it is likely to be greater at larger currents.)
448	55° (± 2°)	Read the value from the graph.
449	700 Ω	• total resistance in circuit = (500 + R)Ω, where R is the resistance of the variable resistor • current = 12/(500+R) = 0.01; hence 12 = 5 + 0.01 R
450	0.8 V	Read from intercept on graph.
451	20 mA	Read from intercept on graph.
452	C	The product of VI at **C** is clearly greater than this product at **B**. The power output at **A** and **D** are both zero.

Capacitors and diodes

453	C	Only these two points give the steadily increasing voltage required.
454	500 μF	Use τ = RC
455		*(graph: voltage across AB vs time(sec), showing exponential decay from 12 at t=0 down towards 0 by t=20)*
456		• the diode will allow through the positive pulses of the AC • between these there will be zero voltage • the graph will look like: *(graph: voltage across XY vs time(ms), showing half-wave rectified sine pulses at 10 and 20 ms, with dashed line at peak value of AC)* • clearly the average will be less than the dashed line
457	No	The logic of an AND gate requires two HIGH inputs.
458	Lit	This will ensure two HIGH inputs.

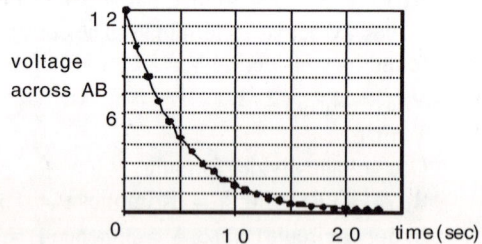

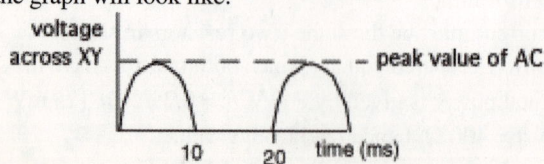

Answers

459	B	• τ = RC; time constant = 10.8 s
		• after one time constant, capacitor will be 63% charged
460		• the point X will be at 0 logic state; this will give the AND gate inputs of 1 0
		• this will ensure a 0 output at 0
461	24.2 s	τ = RC
462		• when switch is closed, voltage of charged capacitor is connected to point X
		• the capacitor then starts to discharge
		• the time it takes to discharge depends on the time constant
		• eventually the capacitor is fully discharged
		• this means a voltage of 0 at the point X
463	A	normal capacitor charging curve, time constant around 1 sec.
464		• the voltage across the capacitor equals the voltage across the battery, therefore the voltage across the resistor is zero; hence I = 0 (from V=IR)
		OR
		• when fully charged, no more charge flows on to the capacitor, therefore there is no current through the capacitor
		• since resistor is in series with capacitor, there is no current through it either
465	C	normal discharge curve for capacitor, but with time constant of 0.2 sec.
466		• this is an example of *half-wave* rectification (rather than *full-wave*)
		• in this case the negative parts of the AC cycle are simply suppressed (due to the diode being reverse biased)
		• the sketches below illustrate the second point:

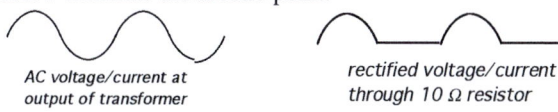

AC voltage/current at output of transformer

rectified voltage/current through 10 Ω resistor

467	0.1 s	use τ = RC
468		

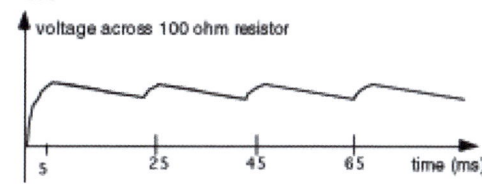

voltage across 100 ohm resistor

time (ms)

469	1 ms	use τ = RC
470	A	τ must be >> than the output signal period; there would be significant ripple for the 0.1 ms signal (or the other longer options)
471	100 µF	use τ = RC
472	D	classic smoothing circuit; capacitor stores up charge during positive gradient positive voltage sections, and releases charge during the rest of the cycle. Note that the time constant (30 ms) is longer than the gap between the lumps of the rectified voltage.
473	1.2 × 10⁻³ F	from τ = RC
474	D	Across the resistor the voltage is high at first, when the current is high, but drops off as the capacitor fills up with charge.
475		

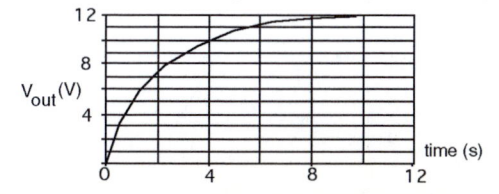

V_{out} (V) vs time (s)

Answers

476	The 0.5 kΩ is part of an open circuit and cannot have current flowing through it. The 1 kΩ resistor is in series with the charged resistor and must have the same current through it as the capacitor. Since the capacitor is fully charged, the current through it is zero.	
477	C	The time constant for the discharge is half that of the charging time constant (from $\tau = RC$).
478	• $\tau = RC = 0.5$ ms • this will be much too short to smooth a rectified AC supply • RC is much shorter than the time between rectified pulses of 10 ms or 20 ms.	
479	2.7 s	The current will drop by a factor of e^{-1} in one time constant. On this graph, this is dropping from 1.2 to 0.44 A
480	27 kΩ	Time constant = RC; substitute the answer to the previous question and C = 100 μF
481	C	The diode will block the negative flowing part of the AC cycle.
482	D	The capacitor must be in parallel with the rectified output.
483	A	The smoothest output will be the RC combination with the largest time constant; this will simply be the largest capacitor.
484	9870 μF	Use $\tau = RC$
485	Increase	The capacitor would have to discharge at a slower rate, so the time constant would need to be larger. Increasing R would achieve this effect.
486	C	Although the charging *rate* would be the same, the voltage would reach 8.0 V more quickly, simply because it is 'closer' to 10 V.
487	The maximum value of the voltage will be +5 V and the minimum will be -5 V.	
488	*graph: voltage (V) vs time (ms), showing rectified half-wave pulses at 0, 50, 100 ms*	
489	*graph: voltage (V) vs time (ms), showing smoothed rippled output at 0, 50, 100 ms*	

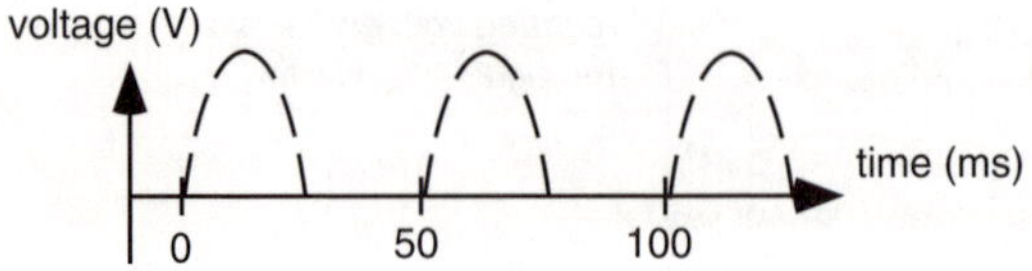

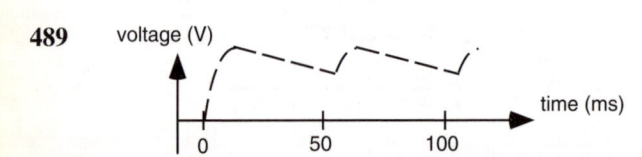

Amplification

490 *graph: voltage out (V) from -2 to 2 vs voltage in (mV) from -4 to 4, linear through origin*

491 *graph: output voltage (V) from -1.0 to 1.0 vs time (ms), trapezoidal waveform with transitions near 2 and 6 ms*

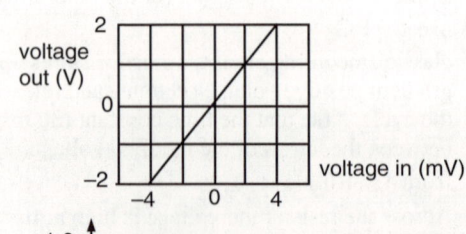

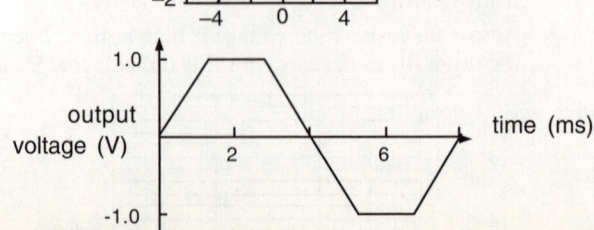

Answers

492 × -5 — The gain of the amplifier is given by the gradient of the graph. The minus sign indicates that it is inverting.

493

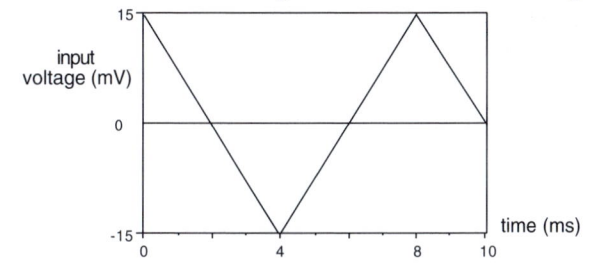

[Graph: input voltage (mV) vs time (ms), triangular wave between -15 and 15, period 8 ms]

494 When input voltage is low, output is high – the condition for a NOT gate.

495 × 15 — Gradient of the graph

496 ± 0.2 V_{peak} — Greater will move the amplifier out of the linear region.

497 × 6 — Gradient of the graph

498/9

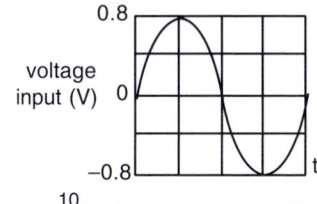

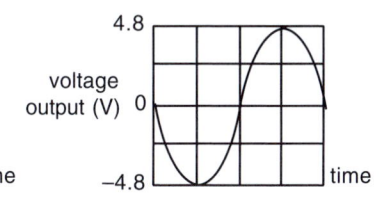

[Graphs: voltage input (V) sine wave between -0.8 and 0.8; voltage output (V) inverted sine wave between -4.8 and 4.8]

500

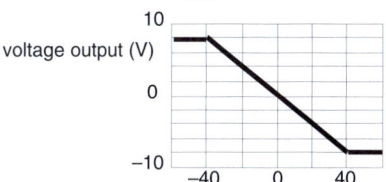

[Graph: voltage output (V) vs voltage input (mV), transfer characteristic from +10 V (at -40 mV) sloping down to -10 V (at +40 mV)]

501 × 200 (–ve) — Gradient of graph

502 80 mV — Greater will move the amplifier out of the linear region.

503

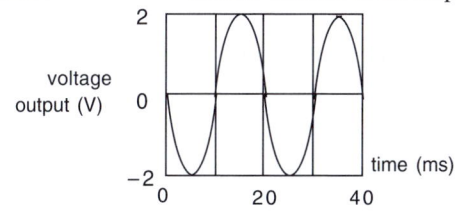

[Graph: voltage output (V) vs time (ms), sine wave between -2 and 2, two cycles in 40 ms]

504 B — Product of gains must equal 1500 (= 3.6/0.0024)

505 × 20 — = 48/2.4

506 1.4 m s^{-1} — Draw graph and interpolate

507 5.9 V — $V_{p-p} = V_{RMS} \times 2\sqrt{2}$

508 8.5 mV — The linear region is 12 mV peak – this is $12/\sqrt{2}$ V RMS.

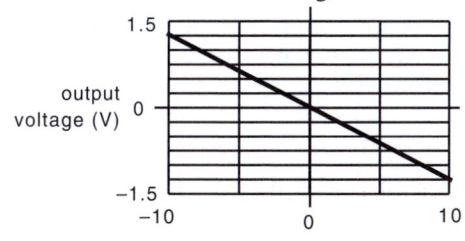

509

[Graph: output voltage (V) vs input, linear from (-10, 1.5) to (10, -1.5)]

Answers

510	× 12	gradient of graph in this region (note: this amplifier requires an input signal with some DC added to ensure that the AC varies in the 'amplifying' range.)
511	C	amplifier non-inverting (note: only AC component is shown.)
512		

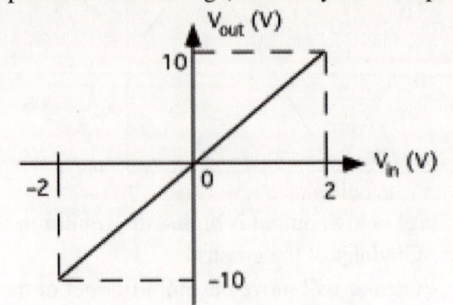

V_{out} (V) vs V_{in} (V): line through origin passing through (2, 10) and (−2, −10).

513	× 5	
514	13 V	23 mV (RMS) is 65 mV (p–p); then gain of 200.
515		

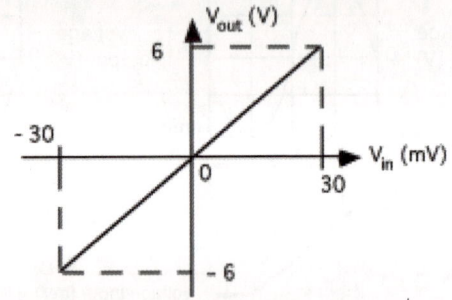

V_{out} (V) vs V_{in} (mV): line through origin passing through (30, 6) and (−30, −6).

516	18 mV	$V_{pp} = 50$, $V_p = 25$ therefore $V_{rms} = V_p/\sqrt{2}$
517	× 100	gain = $V_{pp}(out)/V_{pp}(in)$
518	500 Hz	f = 1/T (read T from graph)
519	−90	gradient of graph
520	B	appropriate gain and out of phase with input
521	× 5	Take ratio from peak values on graph.
522	50 Hz	Measure T from graph (20 ms); use f = 1/T
523	0.71 V	$V_p = V_{RMS} \times \sqrt{2}$
524	5 mW	P = VI
525	18, non-inverting	From the gradient of the graph, gradient is positive.
526	0.88 V	$V_{p-p} = V_{RMS} \times 2\sqrt{2}$
527	0.83 ms	T = 1/f
528	BCD	A would produce too much gain.
529	× 50	• input peak voltage is 0.04 V; output peak voltage is 2 V • gain is ratio of these two
530	A	The amplifier is an inverting one (see graphs in previous question); it is linear (there is no distortion evident in the graphs—and anyway distortion is not on the syllabus...)

Logic

531		• the logic is that of a NOR.
532	A = 1; B = 0	Construct a truth table and follow the logic through.

Answers

533 Yes
- final NAND only gives a LOW when both inputs are HIGH
- this occurs when both the buttons are HIGH (AND gate gives a HIGH) and when the pressure switch gives a LOW (no-one standing on it)

534

Input A	Input B	Output X	Output Y
0	0	1	1
0	1	1	0
1	0	0	1

535/6 A
- before the buttons are pressed both outputs are 1
- the bells are silent
- as soon as one button has been pressed, the output of that NAND will be 0
- the bell attached to that NAND will ring
- this feeds back to the other NAND input, holding it at 1
- so you can tell who pressed first – the one whose NAND output is zero

537

A	B	Output of OR	Output of NAND	Output X	Output Y
0	0	0	1	0	0
0	1	1	1	0	1
1	0	1	1	0	1
1	1	1	0	1	0

538 The Output column reads (downwards) 0, 1, 1, 1.
539 Remove final NOT and replace final AND with a NAND. Other possibilities.
540 Output column reads (down) 0, 1, 1, 1. This corresponds to an OR gate.
541 Output column reads (down) 0, 0, 0, 1. This corresponds to an AND gate.
542 AC   Both correspond to a NOR
543 same shape as mass sensor graph
544 there is a can of the correct mass present
545

V_L	V_M	V_{OUT}
1	1	0
1	0	1
0	1	0
0	0	0

546 D   follow the logic through
547 5 V, logic 1   this is the function of a NOT gate
548

lift movement	lift door	warning
0	0	0
0	1	1
1	0	0
1	1	1

549 A   D would also give a 1 (but also gives 1 for lift moving up)
550 The lift is overloaded but the door is shut.
551 B   both switches need to be closed for the light to operate
552 A   either or both switches will turn the light on
553
- the circuits are C and D
- for C, L is only on when $S_1 = 1$ and $S_2 = 0$
- for D, L is only on when $S_1 = 1$ and $S_2 = 0$

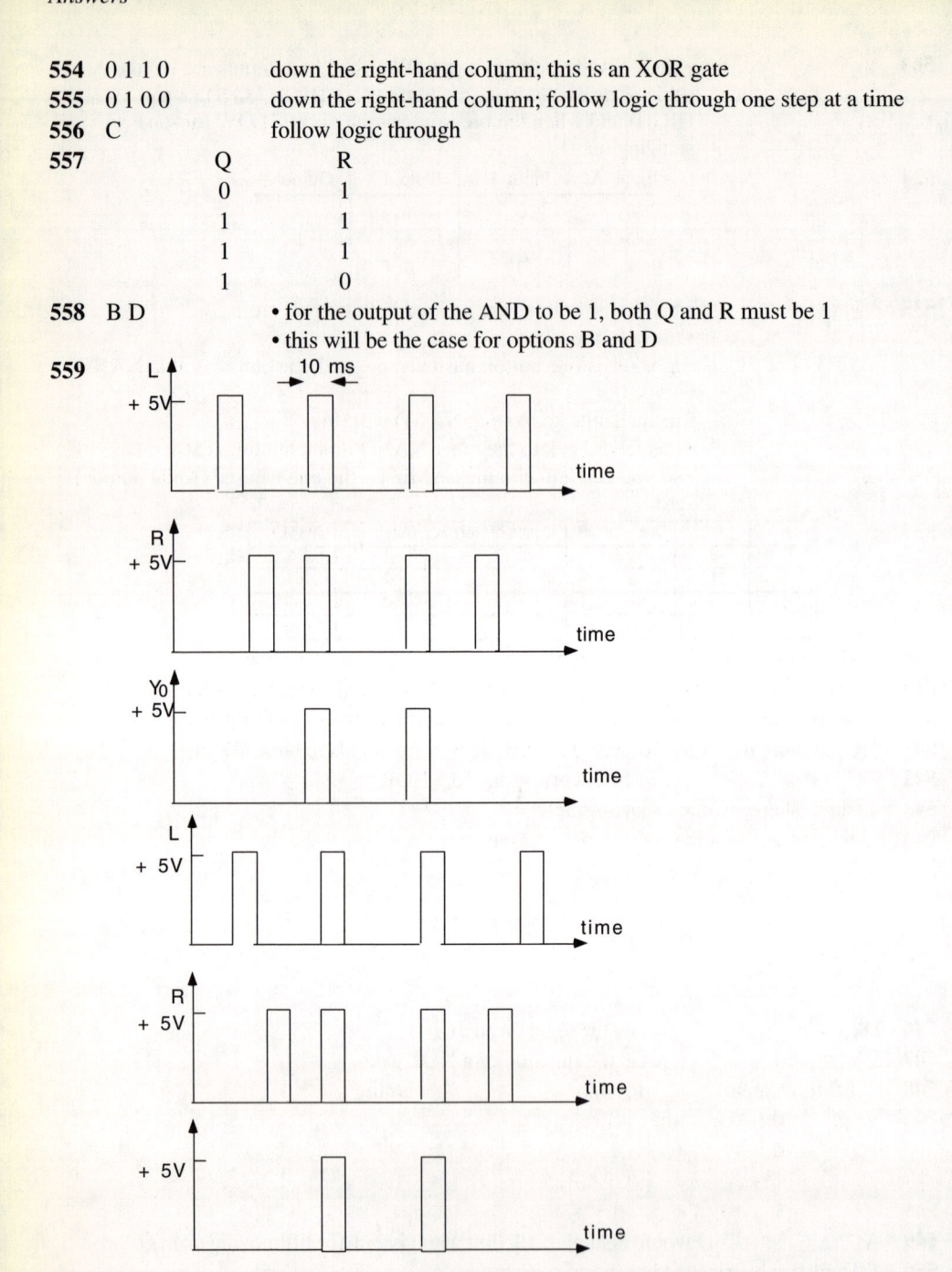

Answers

554 0 1 1 0   down the right-hand column; this is an XOR gate
555 0 1 0 0   down the right-hand column; follow logic through one step at a time
556 C   follow logic through
557

Q	R
0	1
1	1
1	1
1	0

558 B D
- for the output of the AND to be 1, both Q and R must be 1
- this will be the case for options B and D

559

196

560

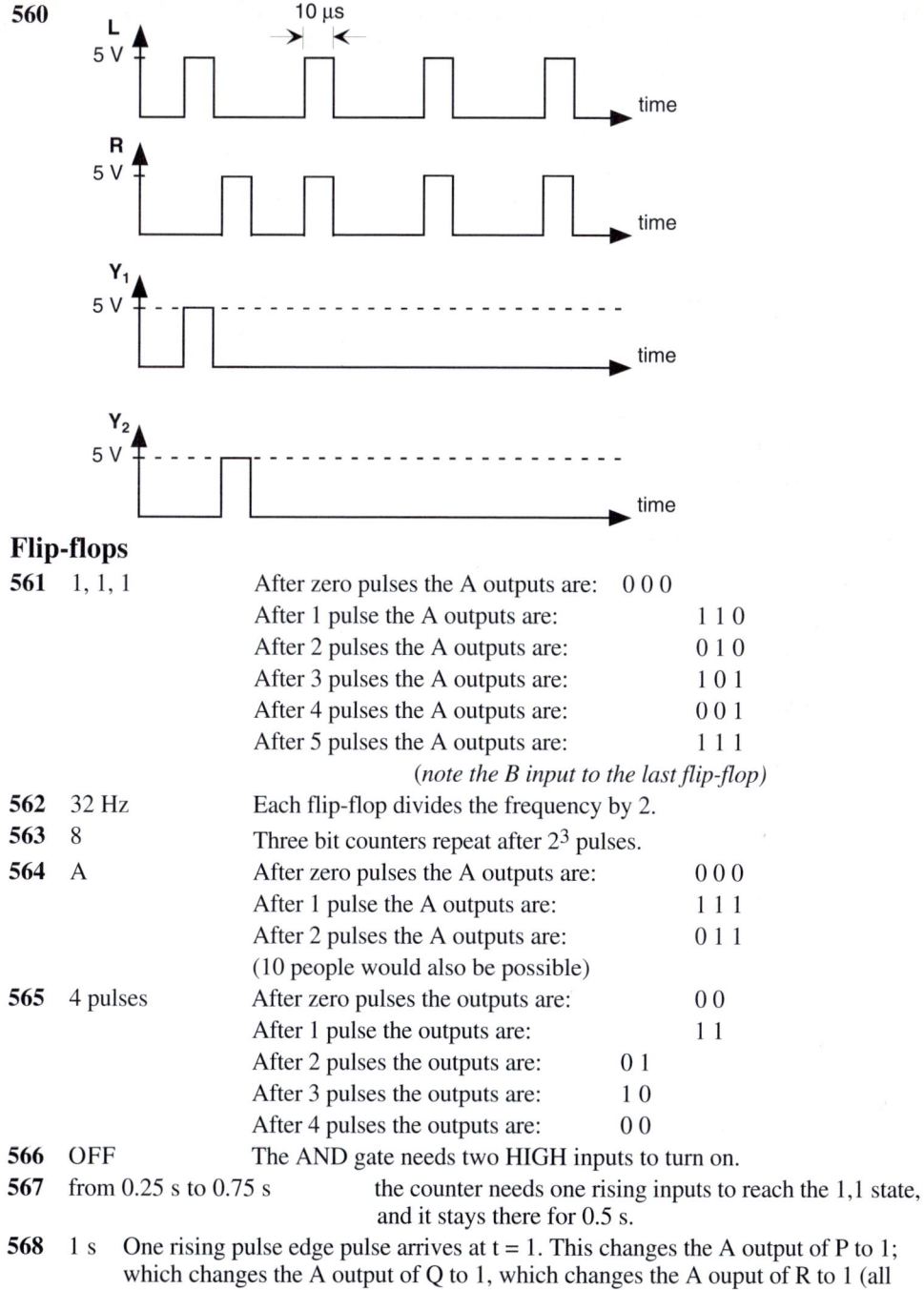

Flip-flops

561 1, 1, 1

After zero pulses the A outputs are:	0 0 0
After 1 pulse the A outputs are:	1 1 0
After 2 pulses the A outputs are:	0 1 0
After 3 pulses the A outputs are:	1 0 1
After 4 pulses the A outputs are:	0 0 1
After 5 pulses the A outputs are:	1 1 1

(*note the B input to the last flip-flop*)

562 32 Hz — Each flip-flop divides the frequency by 2.
563 8 — Three bit counters repeat after 2^3 pulses.
564 A

After zero pulses the A outputs are:	0 0 0
After 1 pulse the A outputs are:	1 1 1
After 2 pulses the A outputs are:	0 1 1

(10 people would also be possible)

565 4 pulses

After zero pulses the outputs are:	0 0
After 1 pulse the outputs are:	1 1
After 2 pulses the outputs are:	0 1
After 3 pulses the outputs are:	1 0
After 4 pulses the outputs are:	0 0

566 OFF — The AND gate needs two HIGH inputs to turn on.
567 from 0.25 s to 0.75 s — the counter needs one rising inputs to reach the 1,1 state, and it stays there for 0.5 s.
568 1 s — One rising pulse edge pulse arrives at t = 1. This changes the A output of P to 1; which changes the A output of Q to 1, which changes the A ouput of R to 1 (all *rising* edge pulses).

Answers

569 5 s

- the AND gate output will change from 0 to 1 with two HIGH inputs
- this occurs when the counter reads $(A_P, A_Q, A_R) = (1, 0, 1)$
- this happens at t = 5 seconds (see table at right)

time	input	A_P	A	A_R
0	0	0	0	0
1	1	1	1	1
2	0	1	1	1
3	1	0	1	1
4	0	0	1	1
5	1	1	0	1

570 8 s

frequency is 0.5 Hz; period is 2 seconds; 4 periods are required

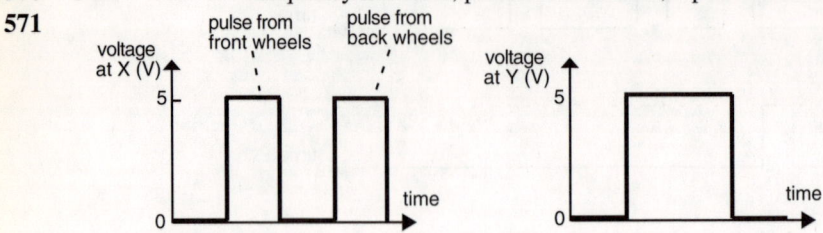

571

- note that each two pulses from a car produces one pulse for the counter

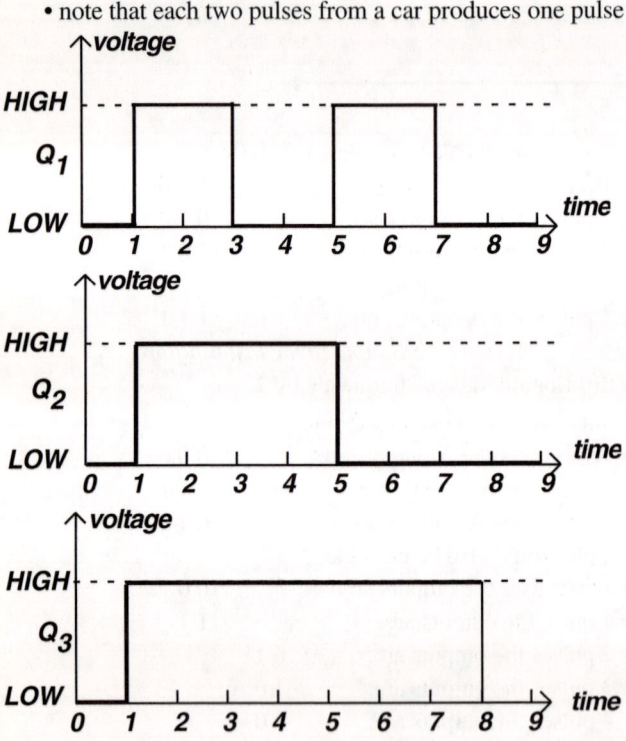

198